Praise for House Rules

"Physical abuse is a terrible thing, but sometimes it seems we diminish the real pain caused by emotional abuse. As Rachel Sontag makes clear in her searing memoir, *House Rules*, emotional abuse can be as devastating, as cruel, as the most severe physical and sexual maltreatment. . . . Almost everyone will see a little bit of himself or herself in Rachel Sontag. Whether in the schoolyard, at home, or in the office, bullies are everywhere. What is remarkable and inspiring is that Sontag emerged from the situation a stronger person."

—*San Francisco Chronicle*

"Sontag's lean writing captures the tension—the feeling of family as prison. Each time an outside observer recognizes her father's manipulative cruelty, the reader feels a little surge of hope. Get out of there, Rachel! Get out!" —*Los Angeles Times*

"Imagine: You grow up in an affluent midwestern suburb. Your parents are professionals—Mom is a social worker; Dad's a respected doctor. You do well in school, are involved in activities, and always finish your chores. You are a Good Kid. But your father records your phone conversations, makes you write apology letters for wronging him, and locks you out of the house if you forget your keys. He wishes you'd never been born, and tells you so. This was Rachel Sontag's life, which she writes about in vivid detail in her memoir, *House Rules*. Sontag finally got free of her father's grip (no thanks to her spinally challenged mother), and made her way to New York, where she eventually received an MFA from the New School. Her story shows just how resilient the human spirit can be." —*Gotham* magazine

"It's a thought-provoking and sensitive book."

—*Sarasota Herald-Tribune*

"Somehow, Sontag rises above the predictable in this gripping, quirky, unusual look back at a childhood that would have ruined adulthood for most people. Sontag's voice remains clear, authentic, and humorous throughout." —*Library Journal*

"Sontag's is a brave account not only of what it's like to take the brunt of an abusive parent's wrath but of what it means to have the courage to leave." —*Publishers Weekly*

"In this brave, hard-won, and gorgeously written memoir, Rachel Sontag lays out the story of her family in prose as tautly strung and delicate as a high-wire. To have endured her painful past is one thing. To have crafted a thing of beauty out of it—well, that is quite another. *House Rules* is a remarkable book that will move readers as surely as it will instruct them in the art of survival."

—Dani Shapiro, author of *Black & White*

"*House Rules* is a fresh and utterly engrossing memoir of domestic unease. With a deadpan eye trained on the complexities of family life, Rachel Sontag has written a father-daughter story full of candor, truth, betrayal, and, ultimately, love."

—Danielle Trussoni, author of *Falling Through the Earth*

"Rachel Sontag recollects in vivid detail what it is to die a slow emotional death then somehow manage to resuscitate herself. These are the hardest stories to tell, the most complex, and I would venture to say, the most universal. Sontag stumbles out of her parents' house on the verge of collapse, but, miraculously, with a very strong pulse." —Alicia Erian, author of *Towelhead*

"As riveting, passionate, and powerful a memoir as any I have read in recent years, it is also noteworthy for the balance and scrupulous self-scrutiny the writer brings to her younger self. The result—harrowing as the story may be—is a literary delight."

—Phillip Lopate, author of *Portrait of My Body*

RACHEL SONTAG

HOUSE RULES

A MEMOIR

AN ecco BOOK

HARPER PERENNIAL

NEW YORK • LONDON • TORONTO • SYDNEY • NEW DELHI • AUCKLAND

HARPER PERENNIAL

A hardcover edition of this book was published in 2008 by Ecco, an imprint of HarperCollins Publishers.

P.S.™ is a trademark of HarperCollins Publishers.

FIRST HARPER PERENNIAL EDITION PUBLISHED 2009.

Designed by Jessica Shatan Heslin/Studio Shatan, Inc.

Library of Congress Cataloging-in-Publication Data is available upon request.

ISBN 978-0-06-134123-6

09 10 11 12 13 ID/QF 10 9 8 7 6 5 4 3 2 1

For my sister, Jennifer Sontag,
and my cousins Jill and Debbie Sontag

THE THINGS WE LOVED: AN INTRODUCTION

At a recent dinner party I sat across from a man who asked what I was writing.

"A book about family dynamics," I said, hoping that would satisfy.

"You mean a book about *your* family dynamics?" he asked.

"Yeah, kind of."

There was a silence, which I mistook for the end of the conversation, but the subject was resurrected in between the salad and the fish.

"So who's the monster?" he asked. "Your dad or your mom?"

"I wouldn't put it that way."

"C'mon, it's got to be one or the other."

"It's more about the way we worked."

His lips folded into a smile.

"Or didn't."

"So, who's the monster?" he asked again.

"I don't know."

"Are they going to read it?"

"It's far from done."

He watched my face. "Which one? Give it up."

"My father," I said. "He plays the leading monster."

Later in the evening, seated at a table covered in crumbs and wine stains, the man asked if I still talked to my father.

I told him I didn't.

"That's sad," he said.

"Let me ask you," he said, digging his elbows into the wooden table, leaning in closer. "Was it really that bad?"

His teeth were purple from the wine. Except for the two of us, the party had moved into the living room. The man stretched out his legs, placing his napkin on the table.

"I think he meant well," I said.

The man crossed his arms.

"But we were terrified of him."

His confident face was now dabbed with flecks of uncertainty: the uncertainty of a man who wants to know more without being held accountable for an appropriate reaction. Very quietly, he asked, "Did he hit you guys?"

"Never."

That took care of that. The uncertainty drained from his face. He smiled, relieved that we wouldn't have to go *there*. He opened his mouth to speak, decided against it. I could tell, by the careful way he rested his eyes on the large piece of art hanging from the wall, that he thought he could help me see that it wasn't all that bad.

"Was there anything you loved about your father?"

"There were things."

The things we loved about Dad were the things we could predict: The way he rubbed behind our necks when he was in a good mood. How he washed his hair with dish detergent in the kitchen sink. His many practical uses for duct tape: mending broken shoes, hanging timelines of Egyptian civilization across the living-room wall, fixing the kitchen floor, taping the thermostat to sixty-eight degrees and the radio dials to NPR. His insistence on reading the subtitles of movies we watched aloud even though we, too, could read.

Dad was a funny man. That's what other people told me. He humored waitresses and cab drivers, security guards and dentists. They gave off-kilter responses to his jokes, as if they knew that something was funny but couldn't quite nail what.

"Just laugh at his jokes and he'll like you," I told my first boyfriend. "He doesn't care if you're a good driver or if you make a pass at me, he's not the overprotective type."

I was wrong about that. He just wasn't overprotective in the typical way, like he wasn't funny in the typical way.

Dad loved pranks. Our carpet was scattered with plastic recreations of vomit and dog shit and rubber chickens.

"Do you want to see a picture of the kids when they were young?" he'd ask a stranger on the plane, pulling a photograph of two chimpanzees from his wallet. "That's Jenny and this one's Rachel."

My sister, Jenny, and I executed the perfect response on a regular basis. We laughed. We rolled our eyes. We swatted at Dad's sleeve.

We were allowed to be funny back but only to a certain extent. One summer day Dad, Jenny, and I were walking the dog through the forest preserve. I must have been in the seventh or eighth grade. We came across a sign that read NO DOGS ALLOWED.

"Dad, you'll have to wait right here," I said, chuckling.

"What did you say?" Dad asked. "What was that, Rachel?"

Something in his voice wasn't right and I immediately wished I hadn't spoken.

Jenny's laughter came from her nose in suffocated fits. Dad looked at her, somewhat bewildered, and I prayed he'd linger there, on Jenny's laughing face.

"Did you say what I think you said, Rachel? Because you're not funny."

Dad's eyes looked as if they'd been overexposed to the sun, but it was not a sunny day.

"You don't understand humor like I do."

He was very protective of his funniness and, in time, we came to gauge not *what* was funny but *when* we could be funny around Dad.

When we gauged wrong, we paid. Dad didn't speak to me for the rest of the day. I apologized to him before I went to bed, told him that I didn't think he'd take it so personally. It was a stupid thing to say, I mumbled. He looked up from what he was reading and I thought for a moment that it was pain I saw in his eyes, but it was always hard to differentiate tiredness from suffering in the face of an adult. He returned to what he was reading and said, "Tell me, Rachel, what's not personal about calling your father a dog?"

ONE

There was a time before. There is always a time before. It was a time we can all look back on with a certain nostalgic affection. Not because things were easy, but because we all knew our place in relation to Dad.

It was before I turned ten. Jenny was seven. We slept in the same bed. We bathed together. Dad referred to us as "the children," and because we were "the children," because there was nothing distinguishing us from each other, we fought on the same team.

Dad appeared to us as stubborn and erratic, but he was our dad and each of us was desperately trying to feel our way to his heart. Mom was her own person. She had a laugh that filled a room. She set up watercolor paints on the kitchen table, clay on the floor, mixed newspaper with water and paste so we could make papier-mâché masks.

We lived in a house with a yard. We had a dog. We traveled frequently. We saw our parents above us, our protectors, the people who turned our lights on in the morning and off again at night so we could sleep. Jenny and I hurt to hear them fighting, to think there might be something wrong with the foundation upon which we built our images. It was normal stuff that concerned us all back then. But things were beginning to change. Mom was losing her footing.

When I was eleven and Jenny was eight, we attempted to smuggle our Barbie dolls across the Mexican border, to vacation with us in Cancún, where they could sunbathe and swim in the bathtub.

Jenny's Barbies were in better condition. She didn't stick their heads in bowls of blue food coloring like I did. She didn't chew their feet off. We married some, divorced others, baptized their babies and threw bat mitzvahs. We traded their clothing and the high-heeled plastic shoes that never quite fit their overly arched feet. We pulled their arms off and taped them back on with Dad's duct tape. We broke their legs so we could build wheelchairs. We gave them names that we'd wanted for ourselves: Brigitte, Kimberly, Tina.

The dolls got as far as O'Hare. In the baggage-check line, Dad caught sight of the circular cookie tin under Jenny's arm. His face soured.

"Ellen. What's in the tin?"

Mom looked at it as if it was an alien object she'd never seen before.

We were standing behind a family of four with a boy and a girl around our age. Bratty-looking, I thought. The girl's fingers were wrapped around the neck of a pink stuffed animal. The boy wore a Hard Rock Cafe shirt that came down to his knees. The dad toted golf clubs. The mom wore heels.

The ticket agent motioned us toward the counter. "How y'all doing?" she asked. No one answered.

"How many bags y'all checking today?" she asked, smiling at Dad.

I watched her thin, frosted lips move automatically. Not very good at reading people, I decided. She didn't seem to realize that something was the matter.

"Sir, how many bags y'all have?"

Probably she wasn't from the South but had flown there several times and enjoyed the sound of the accent.

Dad gestured for the ticket agent to hold on as he waved the three of us out of line. We moved off to the side. Mom stood with her mouth agape, hands on her hips.

Dad examined her as if she were a piece of art he found only slightly interesting.

"What's in the tin, Ellen?"

Mom fumbled with her purse.

The next family stepped up. Kids with yellow headphones stuck on their ears.

"The girls' stuff," Mom said.

"Stuff?" he said. "What kind of stuff?"

Jenny and I knew when it was going to get bad. We could always feel it. Dad was about to launch an attack that Mom could not deflect, and we waited, anxious and excited that it was Mom he was mad at and not us.

"What's in the tin?" he asked.

Mom looked hard at it, as if meditating on the matter could turn the dolls into a stack of *National Geographic* magazines.

"Barbies," Mom said.

"Barbies?" Dad took a step back. "Are you kidding me, Ellen?"

His face went white. His lips curled upward, and if one didn't know his many degrees of anger, it would be easy to mistake his face for amused, which he was not.

The ticket agent looked at us. "Sir, you guys ready for check-in?"

"No," Dad said.

We each took a few more steps away from the counter. Jenny sat down on her duffel.

"I can't believe you've done this. We've talked about this, Ellen."

His words hung heavy in the air, like the powerful stench of a skunk's spray.

Mom dropped her head, defeated, as if she, too, could not believe what she'd done.

No one was actually sure what she'd done, but Dad wasn't going to let it go. Whatever storm was rolling in would knock out at least the next two days of our vacation.

"Steve, this is ridiculous," Mom said.

Ridiculous was something Mom often accused Dad of being, but it was exactly the ridiculousness that kept Mom intrinsically connected to Dad. It made her, momentarily, the object of his attention, albeit through anger, at a time when he was losing interest in her. She got used to Dad's ridiculousness. This was just the way her husband was. And we got used to Dad pulling us out of lines and making scenes.

"Jenny? Rachel? Which one of you couldn't leave the house without your dolls?"

We got nervous. We looked at each other, silently blamed the other.

"We're locking them up," he said, staring at the tin.

"Steve, it's going to cost a lot more money to lock the dolls up at the airport for ten days."

"I don't care. It's the principle."

He looked at his watch and raised his eyebrows. "We've got half an hour. You better find a place to lock those things up."

"Steve, it's not worth it," Mom said.

"Not worth it to send a clear message to the kids?"

Mom proposed bringing the dolls to Cancún and storing them in the closet. Dad stared at the airport floor, shaking his head for a long while, every now and then muttering, "You really don't get it," until Mom had had enough. I could see it in her face, a breakthrough moment in which she really did seem to realize the ridiculousness of it all.

"I get it, Steve," she said, stepping away from the line of people—a good thing, because when Mom got mad, she talked in this slow, strained, British-sounding accent, and people seemed to stare.

"You don't get the importance of consistency, do you?" Dad asked.

In the same way that fake fruit is not meant to be eaten, Dad's questions were not designed to be answered. Jenny knew this and I knew this, but Mom seemed convinced that providing answers was an effective means of communicating with Dad.

"No, Stevie. I understand the importance of consistency very well."

"She's so embarrassing," Jenny said to me.

"We've been doing just fine up until now," Mom said.

"Are you ready to be consistent?" he asked.

I imagined Mom in a cheerleading outfit. Her hair tied up in pigtails, feet in bobby socks, pom-poms in hand. "I-AM-READY!"

Instead, she touched her fingertips to her temples and quietly said, "I'm ready, Steve."

Dad stared hard at the green cursive *MEXICANA* sign, and, for a moment, it seemed he'd lost interest in the situation and the dolls might have a shot at Cancún after all.

"Are you going to lock the dolls up, Ellen?"

"Yes, Steve. I'll lock up the dolls."

"It's going to be very expensive. You're going to spend a hundred dollars to leave those things in a locker because you didn't have the strength to say 'no' to the kids. I mean, who brings a box full of dolls to Mexico?"

Mom moved quickly through the airport, following the signs for the lockers. We followed. If we missed this flight, it would be her fault. She was on a mission, and it wasn't locking up the dolls, it was getting Dad back to that quiet place in his mind where he could enjoy being with her.

"Disgusted by you, Ellen, I'm really disgusted," he said.

"I'm sorry to hear that, Steve. I wish that you weren't. I really wish you weren't."

"You know how much this trip cost, Ellen?"

He addressed the same question to us. "You kids know how much this trip cost?"

I pictured the Barbie dolls coming alive, their limbs taking form, eyelashes fluttering, bodies climbing through the tin.

Dad called himself a feminist, gave his secretary paid maternity leave before it was mandatory, bragged about doing it. He believed Anita Hill, supported a woman's right to choose, later insisted on sending me to a women's college. He worshipped Cokie Roberts, Nina Totenberg, Terry Gross—the many female voices that we heard on NPR. He kept black-and-white photographs of these women in binders, taking them out on occasion to show us the people behind the voices. It was the voices that he liked. That he believed and trusted and felt a certain safety in. These were women without bodies, black-and-white faces. These were women Dad would never have to know, the sexless women that he loved.

The four of us stood around that locker while Mom searched for quarters in her purse. We were used to missing flights; we almost always did. Dad enjoyed the challenge of leaving the house late and racing to the airport to see if we could get on board.

"Twenty minutes," Dad said, looking at his watch.

"We should lock him up while we're at it," I whispered to Jenny.

"Do you think they charge more for locking up people?" she asked.

"I'd pay it, whatever it costs. Dad can play Barbies in the locker while we go to Cancún with Mom."

"He'd probably kill them all," she said.

"No, he'd like them. They can't talk back."

We missed our flight.

We went to get breakfast in the airport cafeteria. Dad wouldn't look at Mom, and Mom wouldn't look at anyone but Dad.

I ordered French toast, since Dad was too mad to take the time to say, "I'm not paying four dollars for a stack of hot white bread." Jenny followed suit. Mom got up from the table and, oblivious to the fact that we were in public, put her hands on her head and let out a shrill that sounded very much like a farm animal being slaughtered. Then she placed her hands on her hips, gave her head a shake, and yelled, "I'm so sick of this, Steve," in a voice that did not belong to her, a voice that sounded like it had run away from the body it belonged to years ago.

I could feel my cheeks turning color and covered my face with my hair. Jenny got under the table. Mom stopped shouting. She put her fist in her mouth. And as if he wasn't hearing what we were forced to hear, Dad sat very calmly at the table, chewing and swallowing his food.

Dad was good at remaining calm. Mom got crazy and Dad

saved us from embarrassment by remaining in control, by pretending, as both Jenny and I learned to do as well, that we did not hear Mom's noise moving through a room, like cold air before a storm.

By the time I began junior high, Mom and Dad fought all the time, often about me. Was I studying correctly? I was getting good grades but was I *really* learning? Dad sat down with me every night after dinner to review my homework. He had specific ways he wanted things done. All homework was to be written down in my notebook and then transferred to a large dry-erase board Dad hung in the kitchen. Each night I was to make an outline of my work for the week, tracking my anticipated dates of completion in a calendar. My time at the library was limited. I was only allowed to go when I had a specific project that required research. Always, Dad requested an outline of how I would use my "unmonitored" time at the library. It was this unmonitored time that would ruin me, Dad was convinced.

When the daughter of our close family friend had her bat mitzvah in Cleveland, Mom and Jenny made the eight-hour drive from Evanston. Dad stayed home with me, and together we worked on a book report. It wasn't a difficult assignment nor was it due for another week. I was furious about not going to Cleveland. I stood by the window as Mom drove away, Jenny and the luggage in back. I remember thinking Dad didn't want to go, that perhaps he wanted to stay home with me. And then, despite my twelve-year-old fury, came an unsettling feeling of flattery.

Dad and I were vegetarians. Jenny and Mom ate chicken and fish. When Dad was out of town Mom would throw a steak into the oven and order us a cheese pizza. It was understood that we would not tell Dad about the steaks or the trips we took to

Hecky's Rib Shack, where Mom double-parked the car, got herself an order of ribs, and then devoured them in the front seat, a napkin tucked into her shirt.

Dad and I often ate together at Pizza Hut while Jenny and Mom had shrimp cocktails at Red Lobster. This was especially the case on vacation. When the four of us dined together, I deferred to Dad to make sure there was no meat tainting our dishes. Pointing to me, he'd inform the waiter: "The two of us are *strict* vegetarians."

I was my father's daughter.

Meanwhile, the seats of Mom's car were stacked high with self-help books. There was one title in particular, *Women Who Love Too Much*, which repelled me in the same way as the menstrual stains I saw in her underwear when I folded laundry. I was almost thirteen. I felt shame in what I saw as Mom's visible weaknesses, and instead clung to the distinct things that made her my mother.

She drove with her right hand on the steering wheel, her left hand combing through her hair. At the stoplight, she'd examine her teeth in the mirror. I used to watch her from the backseat, imagining what she must have been like at my age, noticing the way she signaled her turns too early, then left the signal clicking well past the actual turn. Wondering if she was absentminded or had too much on her mind. Habits that made me think she could get in the car and drive away forever, if that was what she wanted.

She was driving me to band practice the morning of the accident. White metal flew through the red light and into the side of Mom's car, the side where I was sitting. Mom jerked the steering wheel. Brakes screeched. Horns blared. Pain shot through my body.

"Jesus Christ," Mom said. "Are you okay, honey?"

"Yeah," I said. The pain in my ribs made it hard to breathe.

"What a moron," Mom said. "What a fucking moron."

Traffic stopped. People got out of their cars and came toward us.

My door was bashed. I climbed out the other side. Mom walked slowly toward the blond, ponytailed woman in the white Ford. I stood on the sidewalk and watched Mom take control in a way I'd never seen her do.

"What the hell were you thinking?" I heard her shout at the woman.

The woman got out of her car, crying a sort of dazed, hollow cry. She looked around forty. I bet she had no kids. The front of her Ford was totaled. The side of Mom's Cult Vista was badly dented. I stood outside, with my arms around myself, trying not to feel my ribs.

"I feel terrible," the lady said, shaking her head back and forth. "I must have spaced out."

I felt terrible for her, being so clearly in the wrong. It was the worst kind of visible to be, with judgment up ahead. Sirens sounded, police cars pulled up. A fat, redheaded cop got out of his car. A fire truck arrived. Pink flames were placed in the middle of the street. I scanned all four corners of the intersection, highly aware of the scene that was being made, conscious of the kids passing by on their way to school.

"I'm sorry. I'm so sorry. I don't know what I was thinking," said the woman.

She reminded me of Mom, the way she kept saying that.

"You weren't thinking, lady. You weren't thinking at all," Mom said.

There was talk of getting an ambulance. "Just to play it safe," the cop said. I enjoyed hearing this suggestion, as if his concern far exceeded any type of standard accident procedure.

"I don't think I need one," I said.

"I'd like to get you one anyway," said the cop.

He leaned in closer to Mom. "That is one seriously banged-up car. Looks like it probably hit her side real hard." Then, pointing to his stomach, he said, "You've got ribs and stuff in there."

I wondered if he lectured his daughter's dates before he let them take her out.

"We live right down the street," Mom told the cop, motioning with her hand. "Rachel, can you go home and get Dad?"

Glad for something to do, I walked the block and a half home with my hands on my ribs, a hunger-like pain spreading through my abdomen.

Dad was leaving as I approached. Dressed in his baby-blue button-down shirt and a pair of gray slacks, clutching his faux leather briefcase in one hand, coffee cup in the other.

"You're back, Rachus," he said, smiling.

Knowing something Dad didn't know made me want to cry.

"Dad. We got into an accident."

"Where?" he said, the smile vanishing.

"Down the street."

I could feel the hiccups coming on and I didn't want to break down in front of him. I closed my eyes.

"Stop the dramatics, just tell me what happened."

"This lady ran the light while we were crossing Ridge."

"And?"

"And her car came flying into ours."

"And?"

"We got hit."

"Is Mommy okay?" Dad asked.

"She's fine."

"So what are you doing here?"

"What do you mean?"

"What do you mean, *what do I mean?* What are you doing leaving the scene of the accident?"

"I came to tell *you*."

"You don't run away from the scene of a crime. You just don't!"

"Dad, the cops are there."

"You don't do that."

"Okay."

"No. It's not okay. You just don't."

"It was a block away."

"You don't get it. You really don't get it. You LEFT the scene of a crime."

"I was told to come get you."

"Just turned your back and walked away."

I looked at my shoes.

"That's really, really stupid, Rachel."

He stepped onto the front stoop and shut the door, shaking his head. "You never do that again. Got it?"

"Yeah."

"See, I don't think you do."

"I do."

"You do?"

"I do."

"Then what are you still doing here?"

I watched him lock the door then turn away.

"Mom told me to come get you," I said quietly as he passed me on the sidewalk. "Mom's the one you should be mad at."

Dad turned around; it was the last time he looked at me that morning. "Mommy didn't leave the scene of the crime. You did."

He was right in the sense that he was factually correct, and often he was. There was a lesson in this. There were usually lessons to be learned. That was the terrifying thing about Dad.

It was like the lesson of handshaking, which I learned when I was ten. As an adult, Jenny would tell me how her handshake was something men always noticed, on dates and on interviews, how they'd compliment the strength with which she shook their hand. She'd tell me this and I'd agree and then she'd say, "It's things like this that made me love him."

■

It was in the lobby of a fancy hotel in Denver, where we'd accompanied Dad on a business trip. We ran into a doctor friend of Dad's. Although Dad was a doctor, he didn't like most doctors. There was a difference between him and "them," Dad explained. Doctors were arrogant, obsessed with money, lifestyle, appearance. Most doctors didn't value medicine. They spoiled their kids and threw around money and went into private practice to maintain their expensive habits. But Dad seemed to like Toby, who wore a crisp white shirt and a tie, his face blotched red from the sun and his grayish-white hair manicured in a way that made him seem like he might enjoy golf.

"Rachel. Jenny. This is Toby Patterson," Dad said.

And then, with an unusual, almost showy tone of confidence, Dad said, "Toby, this is my wife, Ellen, and my older daughter Rachel and my other daughter Jenny." It was just like the host's introduction at the beginning of *Family Feud*. I stuck my hand out and shook Toby's hand and then his wife's.

After the couple left, the four of us got in the elevator. Dad's eyes were fixed on me and he looked as if he was going to be sick. We stepped out. Instead of heading to our room, we stood in the hallway near the elevator. "Let me have your hand, Rachel," Dad said.

"What?"

"Give me your hand."

I extended my arm.

"Shake my hand."

Hesitating, I opened up my hand and offered it to Dad. With my elbow bent and my fingers pressed together, we shook.

"No," he said. "That's not the way you shake a hand."

"Jenny, give me your hand."

Jenny rolled her eyes and extended her hand, small and pale with five fat stubs.

"C'mon," he said. "You've got to be kidding me!"

Relief filled me like helium. Jenny failed him, too.

"Ellen, they don't know how to shake a hand. Give me your hand, Rachel. Give. Me. Your. Hand."

Again, I extended it, this time flexing all the muscles in my arm. With all my body weight stored in my front leg, I lunged toward Dad, took his hand, and pumped my arm as hard as I could, dropped it, pulled it up, and dropped it again in a wave-like formation that made him so angry I thought he might throw me through the glass window.

Instead, he let go of my hand and very calmly said, "No. No, that's too hard. Give me your hand."

I gave him my hand again.

People had gathered near the elevator, observing the spectacle this lesson had become.

"Feel the way I shake your hand," Dad said, looking me hard

in the eyes, shaking in what I supposed was the proper manner. "Got it, Rachel? Now that is a confident handshake."

We practiced the confident handshake a couple of times, standing out in the hall, Dad introducing himself as Toby. Jenny's problem was lack of eye contact and mine was finding the balance. I was either too strong or too weak for what Dad considered right.

The next day, Dad spotted Toby and his wife in the hotel restaurant. "Look who's here," he said. "Now go over and shake their hands properly."

"Dad, they won't even remember," I said. "They didn't even notice that we did it wrong."

"You don't think they noticed, Rachel? Oh, they did, they did. Everyone remembers a poor handshake. Go over there and introduce yourselves the right way. You're lucky you have another chance. Apologize and give them a good shake."

Jenny and I approached the Pattersons, Jenny standing a little bit behind me, saying nothing. I took the lead and shook their hands but we did not apologize. I could feel Dad watching from a distance, and for a quick second, it wasn't Dad I was angry with, but Toby, for laughing so gregariously, for giving Dad that congratulatory look older men with grown-up kids give younger men with children, as if to say, *Way to teach those kids what matters.*

■

Back at the scene of the accident, insurance information had been exchanged and tow trucks came to take the cars away. The blond lady left. Dad went to work. Mom and I went to the emergency room. A nurse uncapped needles on the table by my bed. Mom left the room, since she was known to faint at the sight of blood. After my blood was drawn she came back holding a can of Sprite. We both watched the liquid from the IV drip into my arm as we waited for the doctor.

Pulling a chair up to the bed, as if he was going to read me a story, the doctor said, "Well, Rachel, you're in for a little more than you bargained for. Your kidneys are bruised pretty badly."

His words were overly pronounced, like a speech therapist's, and I could tell by the thoughtful way he was looking from me to Mom that he was the kind of doctor who took his wife out for fancy dinners. Who retold the family friends, over rounds of scotch, the story of how they first met. He was the father who knew the names of all the girls on his daughter's tennis team.

The nurse put X-ray armor over my chest. I lay flat on my back under what appeared to be a huge silver doughnut, which would move over my entire body as it X-rayed my ribs and kidneys. She prepared a full shot of iodine.

"We're right across from you. If you feel any pain at all, which you shouldn't, give a holler," she said.

I wondered what Dad said to his patients before he gave them colonoscopies. Did he touch those big VA veterans he talked about so much, hardened men with colon cancer, who'd been dragged through the trenches and ruined by addiction? Did he tell them not to cry?

Dad was a different kind of doctor than the rest. The doctor who refused to go into private practice or fly first-class or flaunt his doctor status, who turned down jobs at fancy hospitals to work where he worked. Dad was well respected and frequently published, and had seen enough pain in his patients to dismiss the mild, curable aches of his children. It was people in Mom's profession, Dad said, who brought on all the trouble.

They were the sick ones, the ones who thought they could save other people. Those psychiatrists were the ones who needed help. What a waste of time and money, Dad declared. Mom was a school social worker with a degree in art therapy, and Mom laughed it off, saying, "C'mon, Stevie, social work is a very important profession." A couple of times she came down hard, reprimanding Dad for demeaning her life work in front of us kids. But his contempt gnawed away at all of us.

The nurse wiped my arm with antiseptic and gave me the shot. Pain blasted through my shoulder, the way I imagined a bullet might. There was no time to think about whether this pain required screaming. I could hear my own noise echoing through the room. When I opened my eyes, it was to a crowd of other people's concerned faces, and the loud, steady beep of a hospital machine. I tried to breathe. I couldn't.

The nurses' hands came down on my body, one by one, until I was covered in hands. I heard something like the ocean roaring in my ears. Eventually, the machines stopped beeping, and I saw a big, bloated image of Mom dancing above my head. She was in her denim skirt, with a bowl of egg salad, wrapped from head to toe in phone cord. I reached out for the image.

"An allergic reaction to iodine," said the doctor.

Everyone kept on pattering around the room.

"Very rare," said the nurse. "Terrifying."

Hours later, I was released. Mom wheeled me through the hall toward the bathroom, where I passed out on the floor, vomiting into my lap. Mom wheeled me back to the hospital room. I was readmitted for several more hours. The doctor replaced the IVs. Later, Mom took me home.

"Do you think Dad's still mad?" I asked Mom.

"Who knows? He's been in a mood all week."

"But do you think he'll still be mad at me?"

"Rachel, who cares about Dad?" she said.

Both of us did, very much, and it was a relief when he came home from work and headed straight for the couch, where I was lying.

"So you're allergic to iodine," he said, smiling.

It was the type of pride I knew he felt when I broke my arm in third grade and passed out from all the morphine. These types of reactions fascinated Dad.

I told him all about the pain, how immediate and piercing it was. He nodded, as if he was well aware of that perfect pain I was trying so hard to describe. He wasn't angry anymore, not one bit; it seemed he had altogether forgotten about me leaving the scene of the crime.

He took out his medical model—a human-size plastic body with removable parts—and showed me the kidneys.

"Bruised," he said. "That's nothing, just a matter of time."

"Nothing permanent?" I asked.

"C'mon, Rachel, you know the answer to that. But that iodine, now that's something else. That is something else. You could have died! You came close to death, Rachel."

It was my turn then to shake my head, as if I knew that just-before-death feeling intimately. I let him be impressed with how I had picked myself up from the floor in the handicapped bathroom and washed the vomit off my lap with paper towels. Together we watched the late-night news.

That night I fell asleep on the couch, where I remained worthy of Dad's attention. That might have been the last time Dad and I existed together in that space, where I was just his daughter and he was just my father.

TWO

In the summer before my freshman year of high school, Jenny and I memorized every capital of Western and Eastern Europe. Dad duct-taped color-coded maps of Europe to our living-room walls so we would not forget what empire ruled when, occupied what, and conquered whom. Mom staged a mild protest over the duct tape peeling down the plaster, but this was quickly overruled. "How else is this going to sink into their heads?" Dad asked.

Somewhere around four on a Saturday morning in June, a cab arrived to take us to O'Hare. Jenny and I had our jobs. I was in charge of the tickets. Jenny was responsible for the maps and the plastic-coated books that Mom had borrowed from the library.

"What are you going to do with those tickets?" Dad asked from the front seat.

"Hold them," I said.

"You're going to *hold* four passports and round-trip tickets to France in your *hands*?"

"Yes."

"You think that's a good idea, Rachel?"

"Sure."

"You're that confident you won't lose them?"

"I'll put them away, Dad. Just give me a second."

"Give me a second? Did you tell me to *give you a second*? You're holding three other lives in your hands."

"Do you want to hold them?" I asked.

"Didn't I give them to you?"

"Yes?"

"You don't trust yourself?"

"No. I do."

"You trust yourself holding on to all of our passports and tickets?"

"Yes."

"It isn't just your ticket you're responsible for."

"I know that."

"Then why are you still holding them?" Dad's voice was getting louder. He turned away from the driver, whom he'd been engaging in conversation about his travels through North Africa.

"Ellen, do you see what's happening here? Rachel's so confident, that she's going to *hold* our identities in her *hands*, our lives. She's willing to put our entire vacation at stake."

Dad turned to me. "I got everyone a fanny pack for a reason," he said. "Are you concerned about looking cool?"

I said no, I was not concerned about looking cool, though this was not the case. I was fourteen and very concerned with how I looked.

"You afraid that people won't like you?"

"That's not it."

"So terrified that no one will like you?" Dad said.

I wondered if the cab driver thought we were nuts, or if he

was even listening. Dad would tip him well. He always did. He also overtipped maids. He liked to wave the envelope in the air before we left a room and have us guess how much he'd left. I used to guess lower on purpose so Dad would have the glory of telling us otherwise. And Dad, satisfied with the discrepancy between my guess and the actual amount he'd left, would say, "This is probably more than they make in a month. This will feed their whole entire family." And while I admired his generosity, I always wondered what gave Dad the right to decide this maid or that driver was the person he assumed them to be.

Our flight landed late at night. People drowsily gathered their belongings. I buckled my fanny pack around my waist. The plane taxied. Jenny fiddled with her hair, oblivious to the books and maps underneath her seat. I watched her face, the way she twitched her nose, the brown mole stuck on her right cheek. My heart pounded frantically inside my chest, quick and nervous, the way it did when I was sick. Just the thought of Jenny getting it from Dad made me high, especially since Dad was in a close-to-perfect mood, making jokes all around and massaging Mom's hand. Jenny was going to lose his maps. Mom laced up her gym shoes, collected her scattered belongings. Jenny would be responsible for ruining his good mood.

Only once, halfway down the aisle, did I think of mentioning those maps, but instead I let Jenny walk away, feeling sick and delighted by her stupidity, "negligence" as Dad liked to call it. I grabbed Mom's hand, my walk turning into a gallop, as we made our way off the plane.

It was not until we got to the car rental place that Dad asked Jenny for the maps. She reached for the backpack, unzipped it all the way, stuck her hand in, and pulled out nothing. Looking worried, but sure she had them somewhere, she unzipped her fanny pack. I was buzzing inside, feeling guilty, ecstatic, and terrified.

When she finally realized what she'd done, she started heaving the kind of terrible bodily sobs that made me embarrassed for her. We were different in this way. Jenny let it all hang out. I checked out. My mind faded to a calming state of numbness. I simply evaporated, removed myself completely from the situation until the situation had passed. Jenny had no ability to turn herself off, to mentally escape a situation she was physically trapped in. It would become our most distinguishing difference as adults. Jenny was hot-tempered and passionate in the unmistakable way of a child who's been ignored.

"What's the problem?" Dad asked.

I couldn't look at her. I was drowning in guilt, knowing the force of what was coming and that it was clearly coming after her. For once Jenny was to blame, and mixed with my guilt was relief that I was off the hook. It must have been the relief that Mom and Jenny felt constantly, since Dad was always on my case. By this point, each of us had quietly accepted that our freedom was determined by the other's entrapment.

"Jenny, what's going on?" Mom asked.

Jenny couldn't speak.

"Where are the maps?" Dad asked. "Don't even tell me."

"Hold on a minute," Mom interrupted. "Jenny, when was the last time you had the maps? Let's think about this."

Like thinking about this was going to get those maps anywhere other than under that seat; Mom was such a social worker.

"They're not in your bag," Dad said. "You have got to be kidding me."

Jenny sucked in air. "I'm sorry," she said through tear-drenched lips.

"That's our whole trip right there."

I started to feel bad for her, really bad. She hadn't even realized the library books were down there too. Mom would pay for that mistake; three hardcover library books and Mom didn't have much money. I wished I'd grabbed the books.

"They can't be trusted, Ellen . . . not at all. They can't be trusted with anything. You with the tickets," he said, pointing to me, "and you with the maps. I don't even want to be here with you two."

In time, he began talking to Mom, and soon he was talking to me. But he ignored Jenny completely, his silence a reminder of what she had done. Why his silence upset her so much, I wasn't sure. When Dad was mad at me I prayed for silence.

I admit that I basked in the aftermath of Dad's affection that night. I didn't leave his side. The next day we started out for the Eiffel Tower in our rental car. Jenny got off sooner than expected thanks to Mom, who, on the way to the Eiffel Tower, stopped at an ice cream vendor and ordered four ice cream cones without first asking the price. They cost an absurd amount of money, which she didn't have on her, and because the ice cream was already scooped, Dad had to shell out the money.

"Such a gross waste, Ellen," Dad muttered, refusing to eat his ice cream. He was right, of course, but it was a fierce price we had to pay for his rightness.

Our lives were becoming a series of rewards and punishments. Years later, we'd reminisce about occasions with a certain nostalgia, not because of the occasion itself but because all of us still remember the intricate details of whom Dad was mad at and why. Which one of us was on his good side, and how the other two resented her.

As we stood looking over the city from the top of the Eiffel Tower, Dad started joking around about the scene in *European Vacation* when Chevy Chase tosses the dog from the platform where we stood. Dad made motions with his arms, pretending to toss an imaginary dog over the railing, and the three of us laughed with everything we had.

"He's back to normal," I whispered to Mom.

"Well, I shouldn't have bought those ice cream cones," she said.

The next day, it was my hair. We were standing in the dank, windowless lobby of a small pension, debating whether or not to take a run-down room, since all the youth hostels were full.

"Are you going to put your hair up?" Dad asked.

"I guess."

"What do you mean, you guess? Do you know what you look like?"

Dad leaned up against the wall, holding his suitcase.

"Steve. Let's make a decision on the room before we get into this," Mom said.

"Get into what?"

"The hair."

"There's nothing to get into, Ellen. Look at her!"

"I agree. She needs a trim."

"She's fourteen and she's got hair hanging in her eyes like a . . . like a cheap girl. It's cheap!"

"Dad, I'm not fourteen until August."

"I'm going to go pay for the room before we lose it, Steve," Mom said.

"Hold on a minute. What about the hair?"

"Well, she's got two options. She wears it back or she gets it cut," Mom said.

"I'll get it cut!"

"Are you yelling, Rachel?" Dad asked.

It was almost physically impossible for me to refrain from talking back to Dad, and this habit would get worse as I got older. I answered questions I knew were meant to be rhetorical, I laughed at what was not meant to be funny.

"Are you yelling at me Rachel?"

"No."

"Because you have nothing to yell about. You're fourteen years old and in Paris. You know how old I was when I left Chicago for the first time? Eighteen years old, Rachel, before I left the country. And here you are, in Paris, walking around with hair in your eyes and your shoes untied, looking cheap because you think that people will like you a little bit more. Am I right?"

"I'm thirteen, Dad."

"I used to know girls like you," he said. "Girls who hated themselves."

That same day, after we checked into our room, my hair was cut to my ears. When the barber was done, he took out a razor and buzzed around the nape of my neck. I looked at my reflection in the mirror, relieved by my own ugliness.

One night after the haircut, Dad and I took a walk. The streets were clean and wet after a day of rain. It was humid and we were both in shorts. I had a bandana wrapped around my head, covering the remains of my hair.

"This kind of walking will keep you alive for five, ten years longer than expected," Dad said.

He was walking fast and I had a hard time keeping up, but

I did, because I wanted to remain in his good graces. He was once again mad at Mom. She did everything wrong, from buying that expensive ice cream to trying too hard to speak French in public and sounding like an idiot. Jenny, by default of her invisibility, was stuck with Mom.

"They should outlaw escalators," I said.

"Absolutely," Dad agreed. "That's the way to lose weight."

On nights like that, it almost felt like Dad admired me, like I reminded him of himself when he was a boy. We walked faster through the streets, looking at the people, talking about the benefits of bikes in Holland.

"Remember when you took a picture of those punks in the square?" I asked. "The big guy with the huge Mohawk in the Sex Pistols shirt?"

"Of course I remember that. Thought that guy was gonna kill me," Dad said. Dad usually got annoyed when I asked if he remembered this and that, but that night he let it slide.

"Remember how he started yelling at you to give him your roll of film?"

"Yup . . . But I tell you, Rachel, it's worth it to have those things on camera. It's worth it to have moments like that."

"Oh, I know," I replied. "I'm just glad you didn't get killed."

We stopped at the bakery for a chocolate croissant and a loaf of sourdough bread. Dad placed several French coins on the counter and then we stood outside and ate our pastries. Dad put his hand on my neck, no longer buried in a mass of long hair, and we made our way back to the pension.

Men smoking cigarettes lined the streets. Small clusters of long-legged women gathered at the corners. Red, Dad explained, was the color of prostitutes at work. On occasion, the hookers motioned at Dad, who gestured "no" with his face and his hands. Dad, who'd been everywhere in the world, knew how to ward off hookers. I was proud of this skill. Proud that he would bring me here, to a world that defied and repulsed, in its hard and real sadness. The hookers were neither young nor pretty, like they were in the movies and in my imagination, and I was humiliated by their interest in Dad. Wasn't it obvious that he was walking with his daughter?

A bearded man leaned up against a sculpture. The exchange happened so quickly I didn't even catch it. Dad snapped his fingers at the man, and the man turned away.

"C'mon, Rachel. C'mon," Dad said. We turned onto our street. Dad didn't speak. I didn't either.

As we walked up the steps of our pension, Dad said, "That man asked me how much you cost."

I could feel my heart stuck in my throat. Dad just shook his head at the ground, clearly upset. At least it had happened after the haircut. It would have been different if I'd still had hair. Dad would have reacted differently. We both knew it.

"Pervert," Dad muttered.

I had never heard him say that word.

"Really sick, that man," Dad said, reaching for his keys.

Inside the pension, Mom and Jenny were already asleep. I walked down the hall to the bathroom, looked in the mirror. I was ugly with that haircut and eyebrows forest-thick, too ugly for it to have been my fault.

I could count the nights I felt protected by Dad. That was one of them, and, just before we left for Europe, there'd been another. Dad had yelled across the yard at our reckless neighbor, Peter, saying if he ever set foot on his lawn or a hand on his daughter, Dad would knock his teeth out. I remember how moved I was hearing Dad speak that way, for me.

That night in Paris was the last time I felt protected. I had turned an age where men saw me. Worse, Dad could see men seeing me. In his eyes, once I was aware of my effect on men, I had to take responsibility. I had to keep myself unnoticeable. Whether it was the length of my hair, my nails, my skirt, the way I'd shake a man's hand, or the tone in which I answered his question, Dad was disturbed, almost repulsed by the change. He was unprepared for me to become anything other than the girl he could once protect.

Mom and Jenny had an advantage with Dad. They needed him, and he was kinder to them because of it. When Jenny had her meltdown over the lost maps, her crying was her begging, her blatant request to be forgiven, and a return to the place where she'd stay a little girl for as long as Dad would let her.

Later, that same trip, when we reached Hungary, I tried to impress Dad by exchanging my U.S. dollar bills for money on the black market. I'd seen Dad do it all summer, and I'd heard him explain in each Communist country we went to exactly why it was so good for those people to have our U.S. dollar bills, and better for us not to have to pay bank fees or abide by the current exchange rate.

As the four of us walked through Budapest, I noticed a man holding a wad of bills, and dropped back. I exchanged my money and caught up with Dad. I didn't know what to say when he took the bills, held them up to the sunshine, and shook his head.

"You've got to be kidding me, Rachel. Counterfeit. You fell for the oldest trick in the book."

He gave me back the bills.

Mom said something about it being an inexpensive lesson to learn, but I was thinking about the man, the smiling face of a liar, and how I'd lost not money but Dad's respect, and how it would have been such a different night if I had won.

Competition was born, something that further isolated me from Mom and Jenny and brought me closer to Dad. It was probably there all along, but it was heightened in my getting older, my enjoyment of the world Dad exposed us to, my desire to get away from him and simultaneously be protected by him.

Like players in a game of poker, Mom and Jenny knew the rules but didn't have the hand to play, and so they sat back as Dad and I put more and more of ourselves into outdoing the other. I didn't see it then, how much I participated, how frustrating it was for Mom, how boring for Jenny.

THREE

We returned from Europe to a flooded basement, a trail of potato chips on the floor, and trash cans filled with empty beer bottles. Mom and Dad had left our sixteen-year-old neighbor, Krissy, with a set of house keys to take in our mail. After investigating all the rooms in the house and assessing the damage, Dad stood by the door, waiting for Mom.

"Let's go, Ellen."

It was after midnight. Jenny and I stood on a stool in the bathroom, in our pajamas, looking out the window at Mom and Dad on the neighbor's porch. Soon, Krissy and her parents were seated in our living room. Jenny and I stood eavesdropping in the second-floor stairwell, as an apologetic Krissy explained the mess in the basement.

Every couple of minutes, Dad would say, "Do you have any idea what would happen if she was my kid? I mean do you have any idea what I would do to her?"

The question seemed to linger, and for a long while nobody answered. Then Krissy's mom said, "Steve, we are Krissy's parents and she will be disciplined."

Dad wasn't satisfied. He wanted to know how exactly they planned to punish her, and he was not shy about asking.

Over and over, until they got up and left, Dad said, "Tell me how. I want to know exactly how she's going to pay for the fun she's had. I want to know exactly what her punishment will be."

Those neighbors never seemed to be the same around us after that. They would still wave from across the street or ask us how we were. We said okay, we always said okay, but I could tell by the politeness in their asking that they'd seen that part of Dad that made them think otherwise.

■

Summer turned to fall. It was the start of my freshman year in high school. My grandma Dot went in and out of the hospital. For weeks, she lay in bed with tubes up her nose, watching *The Phil Donahue Show*. Grandma and I were especially close. She taught me piano. She made homemade eggplant salad, served it on rye bread. She told me stories of her brothers, of growing up in Minnesota, of her father.

In the middle of the night, Dad sent Mom to wake me. In my pajamas, I sat across from them in the living room. I was sure Grandma had died and I remember deciding to stay strong when Dad told me. It made sense I should know before Jenny. She was younger, more emotional. Dad was red in the face.

"What did you say to her?" he asked. His elbows rested in his lap.

"What do you mean?"

"You spent a good half hour alone in that hospital room. What did you talk about?"

"I don't know, Dad."

"What do you mean, you don't know? You know. You know exactly what you said to her."

By the way his eyes studied my face, I couldn't tell if he

wanted to tackle me to the ground and beat me to a pulp or simply make me disappear. But it came as a relief when I asked if something was wrong with Grandma and Dad said, "No, something's wrong with you."

I assumed that meant Grandma was still alive and I was grateful for that.

At a time when I was coming to regard Mom as a child, Grandma was my answer to every school assignment that required the naming of a role model. She'd married at thirty-eight, which was unheard of for a woman in that day. She'd received a degree in piano from Juilliard. She'd been widowed and lived alone for years before I was born. Looking at Grandma's hands, I wondered how her heart hadn't cracked from loneliness in all the years that she'd been going to sleep and waking up with no one by her side.

"Tell me what you talked about," Dad said.

"School and stuff," I said.

"What kind of stuff?"

"Play auditions."

"Play auditions," he said, smiling, as if the suggestion was somewhat comic. "You talked about me, Rachel."

"No. I didn't."

"To my own mother?"

"We didn't talk about you, Dad."

I'd been glad to get the visit over with. It depressed me to see her there. How slowly she'd talked, how I only half-listened, distracted by the blue-faced woman Grandma shared a room with, a woman with no teeth in her mouth or flowers by her bedside.

"She could die," he said.

"I get that."

Mom looked exhausted. “I think what Dad’s saying is what happens between us as a family is private business.”

“Is that what *I’m* saying, Ellen?”

“What we’re saying,” she said.

“I didn’t talk to Grandma about you.”

“Then why is she getting worse?” he asked.

“Is she getting worse?”

The question seemed to irritate him.

“She’s gotten worse since you went to see her. You don’t get it, Rachel. You’re going to kill her with the things you say.”

“This is crazy,” I said, appealing to Mom.

Mom shook her head. “Steve, I think she understands.”

“She’s sick and weak. Are you trying to kill her?”

“C’mon, Dad.”

“I hold you responsible, Rachel.”

“Steve, she *is* quite sick,” Mom said.

“She’s upset,” he said. “Rachel’s upset her.”

“We did not talk about you, Dad.”

“Negligent and selfish, telling my mother bad things about me. My mother! Makes me wonder what you say to my brother!”

I didn’t talk to Uncle Arthur, who was, as far as I was concerned, Dad’s property. Uncle Arthur had two daughters, Jill and Debbie, and my cousin Debbie and I were inseparable until we drifted apart after junior high.

“The only friend you have is your cousin,” said a girl I went to Hebrew school with, and although I had a couple of other friends, it was true that Debbie was my closest.

We walked to school and back together, every single day. We both took up the baritone horn so that we could sit next to each other in the school band. Uncle Arthur enrolled us in Tae

Kwon Do when I was eight and Debbie was nine, and it was Tae Kwon Do that kept me connected to extended family, long after Dad, Mom, Jenny, and I had become our own separate entity, a floating island in the middle of the ocean.

Tae Kwon Do was the one activity Dad granted me without argument, perhaps because Uncle Arthur picked me up and brought me home. No one seemed to guess that thoughts of Dad inspired my fierce sparring and board-breaking, that Tae Kwon Do was my physical release, my psychological escape from home, the only thing that drained the anger from my body.

Uncle Arthur had some clue that things weren't right, though his knowledge was limited and very much affected by the fact that he was Dad's brother, and like everyone else who knew Dad well, he was both amazed by and terrified of him.

We sat in the living room. "Do you talk to my brother?"

I shook my head. He asked again. And when I said no, he asked again and again until I stopped speaking.

"You make up lies for Debbie and Jill?"

My chest felt like a hardened clump of clay being pierced by something sharper than sadness and duller than anger.

"Get out of here," Dad said, getting up from the couch. "I don't want to see you."

Relieved to be dismissed, I went upstairs and got back into bed.

I wondered how he'd been with Mom, how she'd missed the signs. He couldn't have just turned crazy all of a sudden. I wondered if his own father had infected him with anger. But mostly I wanted to know what he saw in me that caused him to break up inside. Was it in my being born or in my growing up? And if it was that simple, then why didn't Jenny cause a similar storm inside him? It seemed, more than ever, that the mere sight of me repelled Dad.

■

That summer, Dad sent me to wilderness camp. Hiking through the White Mountains in Maine, biking through Nova Scotia. It gave me the chance to get into the mountains and out of my head. It was also the first time I ever talked about what was going on at home, something that caused as much guilt as relief, because it was Dad, after all, who'd made it possible for me to go away. There were twelve kids on the six-week trip and three counselors. I'd never seen mountains that big and I'd never spent such a concentrated period of time getting to know others. The trip itself was challenging. I remember pushing my bike up hills when I could no longer ride and I remember the person in front of me waiting until I reached the top to make sure I was behind. And at night, after we'd cleaned up from dinner, the counselors organized "trust games," which involved us sitting down together and talking about our group dynamic. If we had issues with one another, we were expected to address them in front of everyone, to work them out through conversation. We did this every night, and every night I sat in that circle willing no one to say anything bad about me. Soon enough, a girl from East Brunswick, New Jersey, told the group she had a problem with me. I felt my face turn hot and my heart speed up. She went on to say that I seemed to be holding something back and even though I never complained and helped out and made people laugh, there was something off about me. A void in my chest was beginning to fill with anger—quiet, defeated anger that guaranteed me the right to my hurt, that believed no one could possibly understand that hurt. I was looking down into my lap and the girl, who had been the person I'd connected with best, was still talking. Saying she knew there was more, that we all had our weaknesses, and in fact everyone had been picked on a bit in the group except for me, and she couldn't really explain it but there WAS something in me that wasn't coming out. And then others were agreeing, chiming in with their own theories. Maybe I didn't know how to trust them. Maybe I was so interested in other people's lives and stories because I couldn't talk about my own. I continued to slip away, eyes still fixed in my lap, and then I heard the girl say, "Don't get me wrong, I love Rachel, I just don't know how to reach her."

I had no idea just how much convoluted emotion existed in me. Everything hurt. I could feel my organs, the salt from my eyes and the muscles in my stomach, the voice of that girl and how she'd said "love." At some point, the kids sitting to my right and my left each took a hand, which meant I could no longer cover my face, and after fifteen minutes or so, I looked up and I started to talk. These people, in an isolated moment somewhere in the White Mountains, were the first to know there was something really wrong at home.

That next morning I felt like I was recovering from surgery. Slightly weak, incredibly humbled, cared for, exhausted, and better.

That summer replayed in my mind constantly. The cure I found in being physical, bringing my body over the mountains, my mind over my body. It was the first taste I got of the endurance I'd need to get through life at home.

FOUR

Evanston Township High School was the only public school in the town. Our freshman class entered with over eight hundred students. Four buildings came together to make up one huge campus area. The grounds were sprawling with several cafeterias, a handful of gymnasiums, theaters, and outside courtyards. Classrooms were often far enough away that we needed the full five-minute passing period. Evanston was known for its sports and its theater. I pursued the latter.

The first show I auditioned for was Writers' Showcase, which was student-written, -produced, and -performed in one of the school's many theaters. I read from one of the prepared dramatic monologues, giving my best shot at an unpracticed, tarnished-sounding Southern accent. I'd signed up to read last, so the theater would be emptied of the many others waiting to audition. At the end of the week, a sign hung outside the Little Theater, assigning parts. My name wasn't posted.

I auditioned and got rejected for several more shows. Roles seemed to be reserved for those who were born into the black-turtleneck-wearing, cigarette-smoking, witty-sounding crowd. I didn't make the cut.

I was cast as a tree in the high school's production of *The Wiz*. It was nearly impossible to get rejected from the annual musical. It was the play for nonactors. I had no speaking part but was to appear on stage for a long dance choreographed to personify a tornado. I'd given up on getting "in" with the theater crowd, but at least rehearsals would get me out of the house nearly every afternoon for three straight months.

Dad wasn't happy about the arrangement. He'd worked hard coming up with a working set of household rules, which would have to be amended if I was in the musical. After a few long nights of negotiations, he gave in. I had Mom to thank. She fought fiercely for me, highlighting the importance of extracurricular activities: how they'd been proven to help kids focus on their schoolwork, how things like this would make a difference when I applied for college.

Just before opening night, Dad picked me up from rehearsal on his way home from work. I buckled myself into the front seat, tossed my bag in the back with his briefcase. He looked at me. Something hard washed over his eyes. His face looked pained, as if he had a headache. Then he turned on National Public Radio and kept on driving.

Mom was making dinner when we walked in. Jenny was setting the table.

"I'm going to lie down for a minute," Dad said to Mom. "Why don't you tell Rachel to take that crap off her face?"

The crap was lipstick, which I'd begun wearing that year, applying it after gym class and wiping it off on my walk home from school. It was one of the many things I sneaked. Dad had banned me from wearing rings on my fingers and earrings that dangled below the lobe, or lotions and perfume that he said made me smell desperate. I was usually very careful, removing my jewelry, checking my face in a small compact before I turned the keys to the door of our house.

"I can still see it," Dad said at dinner.

Jenny rolled her eyes, sick of conversations that did not involve her.

"I took it off, Dad."

"Ellen, I can still see the crap on her face."

"Let's see," Mom said, examining my lips with the concentration of a medical professional. "I think they're just red, Steve. I think that the lipstick is off."

"You want to know what I thought when I saw that on your face?"

"Not really," I mumbled.

"Not really?" Dad said. "Well I'm going to tell you. I thought, now there's a girl who really hates herself."

Mom placed her hand on Dad's bowl of soup. "Steve, do you want more?"

"There's a girl who hates herself inside."

"That's a little much, Dad."

"That's a little much? That's nothing. I've seen you before with that stuff on your lips and you know what you look like?"

"A circus clown?" I volunteered.

Jenny retreated into laughter.

"A prostitute," he said.

I considered this for a moment.

"That's right," he said. "A hooker."

"Maybe you just don't like the color," I said. "Maybe something more subdued. A coral or a mauve, something in that family."

"Do you want to get slapped?"

"No," I said.

"Then you better stop being fresh, because this is the truth and whether or not you want to hear it, I'm going to tell you. You know why?"

"Why?"

"Because it's dangerous what you're doing, walking around with that stuff on your face. It's provocative, stupid. You're going to attract the wrong person and then it's going to be all over."

Jenny had stopped laughing but I wished she hadn't. Laughter made it all feel much less real.

"And it's going to be your fault. You walk around looking like that, you pay the price. You got it, Rachel?"

I stood up to get more soup. Just standing over the stove was a much-needed break.

"Why don't you get more soup for all of us," Mom said.

The Wiz opened on Friday night. "It's Sabbath," Dad announced, which meant that if Mom and Jenny planned on going to see the play they'd have to go the next night. Going out on Sabbath was strictly prohibited, although Dad had a rather unreligious style of observing. If we didn't go to synagogue, we stayed home and played board games or watched educational videos.

Even though the exception had been made months before, when I auditioned, Dad was furious that night when I got up in the middle of dinner and walked out of the house to wait for my ride. No one said a word.

Sitting on the bathroom floor before we went on stage, I closed my eyes and let one of the other dancers paint them dark with liner, dust them blue with powder, and thicken the lashes with dark black mascara. Then I covered my lips in red lipstick.

I danced my way through the tornado scene. We were coached to dance our feelings out, to make every little thing inside our bodies come to life. I threw back my head and let the anger sweep me under. Something in me broke, like a pipe exploding, leaking, flooding the stage with all that was raw and disgusting. I wanted to be hit or fucked. I wanted some great force to squash my every organ, to rip me open and extract the ache. I wanted someone else to make me hurt. I wanted a pair of strong, well-defined hands to wrap themselves around my neck, to be kissed deeply, in a way that, for moments, might erase everything. I wanted lips that didn't belong to me to press hard against mine.

Three friends came to see the show. They bought me pink carnations and, after I'd sponged off my makeup and changed clothes, we crammed into someone's father's car and drove downtown. There were three boys for my three friends and, although they weren't officially paired off, there seemed to be something beginning for each of them.

We sat in a booth at Bennigan's. There was either a cake that read CONGRATULATIONS or there wasn't. I no longer remember. But I said there was when I talked to Mom on the payphone appealing for a later curfew.

"It's a special occasion," I said. "And there's a cake."

I could hear Mom discussing it with Dad.

"Your curfew's eleven," she said. "We want you home by then."

I stared at the flying kitchen doors, waiters coming in and out, people clapping hands together for a birthday at the bar.

My mistake was not in asking for the curfew extension but for mumbling an onslaught of hushed but repetitious "motherfuckers" when I did not get my way. And I didn't just say it, I sang it to the tune of "Three Cheers for the Bus Driver." It went something like "You guys are motherfuckers, motherfuckers, motherfuckers, you guys are motherfuckers, motherfuckers you are."

I returned to the booth and announced it was no problem if I came home a little late. Better to be grounded next weekend than to ruin everyone's night, I thought. The cake, if I am not imagining it, had been cut. One of the boys at our table was doing a pretty good impersonation of our old, disgruntled gym teacher.

I was thinking about the kind of plays performed at boarding schools up in the Northeast where some of the rich kids in Evanston were sent. Serious plays, like *'night, Mother*, *The Glass Menagerie*, *Danny and the Deep Blue Sea*. I imagined big glass bowls of fruit punch with scoops of melting sorbet floating on top. The theater clique at boarding school would be easier for me to break into. There'd be minimal contact with Mom and Dad, a letter here and there updating them on my academic progress, acting pursuits, and college applications. I wondered if boarding schools awarded scholarships to people whose parents could pay.

Nathalie touched my arm. I could see Mom slowly finding her way to our table. Dad stood with his hand on the entrance door.

"Okay, honey, it's time to go home," she said.

I made no move.

Dad made his way toward our booth. He was in his sweatpants.

"Ready, Rachel?"

And since he'd asked it like a question, I considered saying "No," but instead reached for my jacket. Dad was staring at his hands, making small, puppet-like motions. "Okay," he said as I buttoned up. "Okay," he said again. "How's everybody doing?"

There was giggling at the table. One of the boys responded with a "Fine."

"Say good-bye," Mom said, an instruction that made me feel like I was seven and leaving the house of another seven-year-old.

"Why'd you have to embarrass me in front of my friends?" I asked when we were almost home.

"Did you hear what she said Ellen? 'In-front-of-my-*friends*!'"

Dad parked the car. The engine stopped rumbling and all of a sudden our voices sounded very loud.

"They're not your friends, Rachel. Didn't you see them laughing at you? It was all over their faces. Those people don't even like you."

When we got inside, Dad held up a mini-tape.

"You want to tell us what you said?" he asked. "Because we have it on tape. We have everything we need."

"I was mad, Dad. I was mad about curfew."

"Let's hear it from your mouth, I want to hear it from your mouth."

"C'mon, Dad."

"Say it, Rachel. Tell us what you called us."

"Motherfuckers," I mumbled.

Dad looked genuinely hurt. Like a little boy who'd lost his mother at the supermarket and was half excited by the prospect of wandering and half terribly lost.

"I called you motherfuckers."

And I really wished I hadn't, not just because I'd pay the price, but because it sounded unusually vulgar coming from my mouth.

■

Monday morning, Mrs. Pollock pulled me out of class. She was the high school guidance counselor, a woman with a head full of red, frizzy hair and a pink lipstick mouth. I had been to her office a few times before, to drop and add classes. She was everything I expected in a guidance counselor: beige walking shoes, thick ankles stuffed into stockings, a knit cover on her tissue box. She clasped her hands on her desk and said, "I talked to your parents this morning."

"You did?"

"Do you know what this is about?" she asked.

I shook my head.

"Your dad had a recording. I believe it was from the answering machine."

I sat up straight in my chair.

"You, calling them a certain name."

"Right," I said. "I think I know it."

"Your dad seems to be very upset. He played me the tape over the phone."

"My dad *called* you?" I asked, more amazed than annoyed.

I'd never heard Dad initiate a phone conversation. He rarely used the phone and, when he did, it was only for a dying patient and only after that patient started dying on our answering machine. Dad treated the phone as if the instrument itself were possessed. His face crumbled every time he heard it ring. If the answering machine was off, he'd let it ring and ring until it stopped, or he'd walk over to the wall and remove the entire device. To ensure I didn't talk when I was home alone, he'd lock the phones in a metal safe before he left the house.

"So, you heard the song?" My body filled with a queasy embarrassment. "I was mad."

"Well, I couldn't hear that much but I got the gist." She looked at me sadly. I wondered if she felt like she had more important things to do than listening to a melodic recording of my swearing.

I stared at the family photos she'd chosen to display in her office and thought about Mom's office and how she probably had similar normal-looking photos of me and Jenny, of her and Dad, in frames on her desk.

Mom and Jenny had come to see *The Wiz* that next night against Dad's wishes.

"What does that tell Rachel?" Dad kept asking. "What do you think that tells Rachel?"

He'd sat at his computer with his back toward us, furious as the three of us walked out that door.

What it told me was that Mom, despite her being inherently weak at heart, still had some concept of right and wrong and she knew it would be wrong to not come see me in *The Wiz*. What it told me was that Mom was still capable of making decisions for herself; something that was about to change drastically in the months to come.

Mrs. Pollock pulled out her eyeglasses.

"I hear you were in a play," she said. "What was the play?"

"The Wiz."

"Oh, *The Wiz,*" she said. "How'd it go?"

"Pretty good," I said.

"Who are you playing?"

"A tree." I stuck my tongue into my gum and laughed.

Mrs. Pollock laughed too. "I hear it's a great production."

Then she paused for what seemed like a very long time and

I thought she was waiting for me to speak, because she was looking straight at me with her head tilted, but I wasn't sure what I could say. I knew that if I opened my mouth I wouldn't be able to stop. Then she folded her hands in her lap. She had big hands. I noticed her nail polish, her wedding band.

"I found the call from your dad very strange," she said.

FIVE

Two boys sat in a parked car outside of Bennigan's, their radio too loud for the neighborhood. The boy in the passenger seat rolled down his window. Northwestern boys, I assumed.

"You girls interested in having a drink?" said the boy in the driver's seat, pointing to the flask between his knees.

My heart sped up. I was very interested in having a drink. It was a Saturday night, my one sleepover of the month, and I was looking for something to happen.

"Let's get food," Nathalie said, "I'm hungry."

I looked at her hard.

"Let's have a drink first," I suggested, in a voice that assumed we were forty-year-old divorcées having a martini on the porch.

Pulling up the silver car-door handle, I held the door for Nathalie.

"You go first," she mumbled, and I did.

There was little conversation. I accepted the flask. I pulled my hair back, closed my eyes, and swallowed as much as I could. The liquor was screeching, and burned my insides. An irritated Nathalie sat next to me with her arms across her chest. She knew better. Of course, she did. She could afford to walk away. She didn't value her time out of the house like I did. I held the flask against my knee, then went for another chug, this time counting to ten while the liquid traveled through me.

"What is this?" I asked the boys, already feeling woozy.

"It's whiskey," Nathalie said.

"How do you know?"

"I recognize the smell."

The boy in the driver seat turned toward Nathalie. "But what kind of whiskey is it?"

Nathalie didn't care.

I took another swig. The car smelled stuffy, of grease and Nathalie's French perfume. I rolled down the window and lit a cigarette.

The boys joined us at a booth inside Bennigan's, where we passed the flask back and forth underneath the table. There was some talking, but I don't remember what was said. It was my first time getting drunk, and even then I was nostalgic about first times, imagining how I might describe this night years after it had passed, irritated that Nathalie wouldn't just let herself go.

"Why'd you drink all that tonight?" Nathalie asked later, as we climbed into her bed.

I closed my eyes. Colors erupted from blackness. "Because I could."

She fell asleep within minutes. I stayed awake, nauseated. The house was quiet except for the ticking of Nat's alarm clock, an irritating repetition that made me even more aware of how awake I was. Rolling onto my stomach, I buried my head in the down of the pillow and thought about what it felt like to be happy, happy in your whole entire body, if only temporarily, how good it felt to be afraid of nothing.

I awoke around three in the morning and threw the covers off my body. Vomit collected in my mouth. I threw up in the hallway on my way to the toilet. I shut the bathroom door behind me, trying to muffle the sounds of my retching.

Mr. Friedman was in the hall with Nathalie.

"What's going on?" I heard him ask. "Is Rachel getting sick?"

I went back to vomiting, pulled my knees into my body, and sat down on the floor.

"Was there drinking tonight?" I heard Mr. Friedman ask. "Was Rachel drinking?"

"Rachel had some whiskey," I heard Nathalie say.

The sink dripped, the Friedmans talked softly on the other side of the door, and then Mr. Friedman knocked and asked if I was okay.

I was sorry. I felt disgusted with myself. Not just for being sick but for being jealous. Angry with Nathalie for taking her ability to tell her father the truth for granted. I was jealous of her comfort. Maybe she wasn't rewarded, but she didn't get punished, either. How easy she had it. The truth was expected of her.

"I'm going to bed, Dad," she said to her father, who said back to her, "Good night, Tata."

An amazing love it seemed, that in the midst of a situation like this, he could call her Tata.

The next morning I cleaned up my mess. Mr. Friedman brought me stain remover and rags and a tall glass of orange juice. I could tell he felt sorry for me by the way he stammered about with a half-smile on his face. When the hallway was clean, Mrs. Friedman asked me to sit with them at the breakfast table.

"We're not going to tell your parents," she said. I felt my body coming back to life.

"We're going to keep this between us," added Mr. Friedman, at which point it occurred to me that they were granting me a favor, cutting me slack. It felt like charity, and I was quick to accept.

"What would have happened if it had been you?" I asked Nathalie.

"What do you mean?" she asked.

"I mean, would they have killed you?"

"They would have been mad."

"Do you think they hate me?"

"No," she said, flipping through TV channels and swigging Diet Coke. "You got drunk, big deal. They'll get over it."

"But do you think that they'll forgive me?"

"Jesus, of course."

"I'm never coming over here again," I said. "I'm so embarrassed, you have no idea!"

And, for a moment, I hated her fiercely for not knowing how badly I felt inside. How lucky she was that she could roll her eyes when her father said, "I love you, Tata." I hated her for not having the same desire to get into the car with those boys and take off, for not understanding how sunken it felt to have nothing to lose.

SIX

The lights were on in the den. The glare from the TV bounced against the window glass. I could see their heads. Mom and Jenny sat on the couch, and Dad on the floor.

I rang the bell. Counted slowly to sixty. Rang the bell again. I'd forgotten my keys. After five minutes of persistent ringing, I walked around the side of the house to the den, where I stood watching them watch TV.

Polar bears walked up and down the screen. *The Wild Wild World of Animals.* I stared at the side of Mom's head, pressed my lips into the window. Had there been no glass pane dividing us, I could have touched her shoulders with my lips. I knocked once more but she didn't look away from the TV. Jenny turned her head a couple of times and stared at me. The soles of Dad's feet were pressed together in the butterfly stretch. He lifted his hands off of his feet, looked at me, and shrugged. Now what? his face seemed to ask. Now what?

I removed my gloves and tried the doors on Mom's car. It was too cold to be outside but I was in the wrong; I'd forgotten the keys. The car doors were locked. I returned to the den window. Now the polar bear was sprawled out on a platform of ice. Dad got up and pulled the shades.

Inside the garage, tall piles of old magazines sat stacked in cardboard boxes. Intricate, thick spiderwebs clung to the corners of the ceiling. Breathing in the familiar smell of mildew and earth, I kicked at a watering can. Then I removed the hose from its holder and smacked it against the floor of the garage, thinking about Mom's hand lingering in her hair as she watched those polar bears move lazily from one spot on the earth to another. I sat down on a pile of records and thought about the bathtub. I liked to fill the tub with water so hot it felt cold, drown my body, and do nothing, think nothing, just close my eyes and listen to the clanking of pipes and the ice-cold breaking of winter outside the window, nothing to see but the steam lines of water's sweat upon the tile walls, nothing to feel but the weight of my own body. It was a perfect peace, a fragile and fleeting separation.

I pulled down my pants and pissed in an old silver tin from Poppin' Fresh Pies. Then I carried the tin out to the backyard and dumped the urine in the grass. Steam rose like small clouds and then disappeared. I returned to the stack of records, feeling something close to invigoration. My chest filled with anger.

I was let in a couple of hours later, when the dog was let out.

"What happened?" Dad asked, his hand casually pressed into the kitchen wall.

"I forgot my key."

"Cold out there, huh?"

He was less angry than I thought he'd be, which meant he probably felt like he had succeeded in teaching me a lesson.

The table was set for dinner. I was hungry and glad they had waited. Mom took out a loaf of bread from the refrigerator and handed it to me.

"I had to go to the bathroom," I said.

"That's a problem when you forget your key," Dad said.

Mom placed bowls of soup on the table.

"That's why I was pounding!"

"Don't do that again," Dad said. "You'll break the window."

"I really had to go," I said, looking only at Mom. It was her I held accountable. Jenny was in no position to have opened the door; I doubted that I would have done it for her either.

Jenny stared into her soup. "Is this what we're having for dinner?" she asked.

"What do you mean, 'Is this what we're having for dinner?'" Dad said.

"I mean is this all we're having?"

"Yes," Dad said. "This is what Mom's made for dinner and this is what we're having."

Once that had been Mom's line: "This is what *I've* made for dinner and this is what we're having."

But something was happening to Mom. In a matter of weeks, she'd grown too fat to wear her jeans. Her normally angular face had rounded out, the life had drained from her cheeks, and her bluish-green eyes hung heavy at the lids, drooping like a watercolor painting. Her blinking was delayed, as if opening and closing her eyes was a task she had to perform, and her expressions stayed frozen on her face like a papier-mâché mask. She moved around the house like a zombie; she slept whenever she could. "I'm very tired, girls," she'd say. "It must be the weather changing."

"I bet you won't forget your keys again," Dad said, passing

me the bread then pulling back the plate when I tried to grab a piece. He passed it again, I grabbed, and he pulled it away. He smiled. I smiled. It was a peace offering. I'd been forgiven. And whether or not he cared, I had forgiven Dad. It was an old-fashioned attempt to teach me a lesson; I expected this from him. But Mom I couldn't forgive. She'd left me outside in the cold.

■

I went looking for answers, digging through Mom's dresser drawers for something to explain her glazed-over eyes, the words that came out of her mouth strained and slow.

The pills were small and white, like aspirin. I found them in her purse. I spilled a couple into the palm of my hand and held them close to my nose, but they had no smell. I wrote down the information on the label.

Lithium: to be taken twice a day, in the morning and at night; to be swallowed whole, not crushed, broken, or chewed; not to be discontinued unless instructed by doctor. The instructing doctor was *Dr. Stephan Sontag.*

"That's what's making her so fucking fat?" Jenny asked when I told her about the lithium. I took the pill in my fingers, the pill I'd taken for proof, though I didn't know what I was trying to prove or to whom, and placed it in Jenny's hand. I think we were both under the impression that if we understood what was happening to Mom, we'd be able to stop it. If the two of us stayed unified, we could woo Mom back to her old self instead of watching her get swallowed up by something beyond her control.

There was a feeling of purpose in gathering the evidence and doing the research. Of searching through old medical encyclopedias Dad kept on his bookshelf. We learned lithium was most frequently used to treat manic depression but could also effectively control or prevent manic episodes in persons with bipolar disorder. We looked up bipolar disorder.

> Bipolar disorder, also known as manic-depressive illness, is a brain disorder that causes unusual shifts in a person's mood, energy, and ability to function. Different from the normal ups and downs that everyone goes through, the symptoms of bipolar disorder are severe. They can result in damaged relationships, poor job or school performance, and even suicide.

I read the description aloud for Jenny, emphasizing the part about damaged relationships, though it was the part about suicide that made me cringe.

> Lithium has also successfully treated schizophrenia in cases where there is a schizophrenic thought disorder accompanied by a change in mood that mimics either mania or depression.

Jenny grabbed the dictionary to look up schizophrenia. But I'd stopped thinking about Mom. She was weak, too weak to leave Dad, but weak was different than sick. For a moment I considered whether Dad was sick. If so, his sickness, defined by a well-manicured description like the ones we were reading about, was diagnosable and treatable. Sick was different than evil. Maybe he was unaware of the things he did. And if that was the case, if he was not conscious of the words he spoke, perhaps he was unaware of the effect they had. If Dad could be cured, all of us could be cured.

■

The more dazed Mom became, the meaner we got. Jenny and I waved our hands in front of her face and made loud clapping noises. We talked to her in slow, hypnotic voices, but Mom didn't seem to register our meanness.

It became increasingly difficult to get her alone, with the exception of time spent in the car, where we had her all to ourselves. And so it was driving back from piano lessons on a Sunday afternoon that I staged the intervention. "California Dreamin'" was playing on the radio. I was up front, and Jenny sat in back.

"I don't think you're manic," I said.

Mom let out a small, unexpected laugh.

She had one hand in her hair and the other on the steering wheel and I could tell I was making her mad. We drove in silence until we got to Howard Street. Then Mom pulled into a lot and parked the car at the cleaners.

"You just don't seem like yourself," I said.

She picked up her purse and slammed the car door.

"You're an idiot," Jenny informed me when Mom was out of sight. "You're just gonna make her more crazy."

Mom returned with several neatly pressed skirts and blouses, got back in the car, and calmly reversed out of the lot.

"Lithium's for schizophrenics," I said under my breath. "You are not a schizophrenic."

We continued down Asbury Street.

"Excuse me," Mom said, her foot pressed down hard on the accelerator. "You cannot tell me what I am!"

We sped down the block at a ridiculous pace, the car screeching to a halt.

"Who the hell do you think you are?" she shouted, turning toward me with her hand up in the air. I ducked down in my seat and covered my head.

"I'm not gonna hit you, Rachel. I'm not gonna hit you."

Mom had tears in her eyes. It was the first time in a while I'd seen her react to anything.

"You don't know a thing," she said, and started driving again.

"I know you're not a schizophrenic," I said, softening my voice to a whisper, realizing the conversation was going nowhere. I wanted her to admit she didn't need to be on drugs, and then I'd rest my case.

"You're a kid, Rachel. You don't know a thing about anything."

"Well, I know about the lithium."

"Oh shut up, Rachel. Just shut up."

"Maybe Dad should take some."

Mom jerked the car over to the side of the street, where the three of us sat parked with the blinkers on. It was something she used to do when Jenny and I were younger and couldn't stop fighting. Patiently she'd wait up front until one of us apologized to the other, then both of us would grow quiet and wait for Mom to ask if we were ready to go home.

"Drive, Mom," Jenny yelled from the back.

"Ellen, please drive," I said.

"Do not call me Ellen."

"Mommy, please drive," I said.

Mom didn't drive. She sat on her hands and stared straight ahead. Her eyes looked worn and tired and no longer wanting to stay awake.

"You don't get it," she said. "There are so many things you just don't get."

She was right.

"I've betrayed Dad's trust," she said. "I've hurt him and he needs me."

"We do, too," I said. "We need you not to be like *this*."

"I've turned Dad into a monster," she said. "I've made him out to be the bad guy. I really have."

"Stop," I said, sorry that I'd brought it up. "Just stop talking."

There were instances that year when Mom resurfaced from her semi-comatose state, the way a strong but momentary sun breaks through thick, looming clouds. One of those instances was the night she walked into my bedroom, holding a box of Milk Bones. She called out for the dog, which was not in my room. I was already in bed.

Shutting my door slightly but not all the way, she put the box on the floor, bent down on her knees, and whispered, "Dad is recording your phone calls. You might want to watch what you say."

I closed my eyes for a moment, in the hopes that I was imagining her standing there in my bedroom.

"Just to be safe," she said.

Just to be safe, one leaves enough time before catching a flight. Just to be safe, one drives around with a spare tire, or a gallon of gas, or a road map.

I sat up in bed. "DO YOU SEE IT'S SICK YOU HAVE TO TELL ME THIS?" I asked.

Mom held her fingers to her lips, shut my bedroom door.

"It's not normal, Mom."

Mom hated it when I made broad, generalized statements like this.

"No family's normal, Rachel."

I picked the plaster off the wall with my fingernail.

"Dads don't tape their children on the phone."

It had the rhythm of "Friends don't let friends drive drunk," and I wished we could have conversations like that instead.

Mom looked at me with pity, as if this was a simple matter that I was complicating.

"I'm trying to help you," she whispered. "I didn't have to come up here and tell you this."

It was true.

"This doesn't happen in other families, Mom."

"Well, you don't have another family. You have us."

"Do you understand that I'm afraid to breathe in this house, Mom? Do you understand I'm afraid to do *anything*?"

"Do you think that you're getting a little paranoid, Rachel?"

"Mom, my conversations are being bugged."

Mom stood up, a look of impatience spreading across her face. "Look, it's an easy thing to get around. You want to make a private call, use a pay phone. Save some quarters. Take the dog for a walk."

"Don't forget you're supposed to be looking for the dog," I said.

Mom smiled at this and, to my surprise, I smiled too. It was late. Both of us were tired. She walked into the hall and again started calling out for the dog.

I wanted to believe Dad's craziness would begin to eat away at her sanity the way it was eating away at mine, that she'd have no choice but to get us out of the house. But Mom had come to accept her life with Dad in the same way a terminally ill patient copes with the everyday effects of sickness.

When I was ten, Dad took us to Morocco. He'd rented a red Ford Fiesta, and I remember pulling off the side of the road in a small desert town. Underneath the blistering orange sky of the Sahara, we drank tea with a Bedouin man who offered his antique teapot in exchange for Mom. Dad laughed. We all did. Dad examined the teapot and shot a glance at Mom, playing it out for a while, bargaining for different things: "I'll take this vase and the teapot, maybe the teapot plus a camel." Back in the car, he asked, "What do you think they'll say at home when I tell them I traded Mommy for an old Moroccan teapot?"

"I think you should have done it," I said, regretting the words as they came out of my mouth, because nothing in the world scared me more than losing Mom.

Mom came back to clear things up in the morning, having discovered that Dad's recording device could only tape incoming calls.

"That's a little relief," she said.

And it was, momentarily. But I knew, by Mom's nonchalant acceptance of things like this, that I was losing her.

SEVEN

It was the winter of my sophomore year. We'd begun to live through Dad the way Floridians live through hurricanes, as normally as we could until the storm hit. It was a harsh winter, a record-breaking cold, and forty-six consecutive days without sun. Dad had duct-taped flannel sleeping bags to the doorways and the windows to insulate us from cold air. Nighttime came quickly. Kids at school talked about daylight hours and depression, shared statistics about suicide in Scandinavia.

Mom's knock against my bedroom door became the summons I'd come to dread. This particular night it was around ten o'clock. She stood with her hand on the knob, staring at the carpet for a long and empty moment. "Dad and I would like to talk to you downstairs."

Chilling words. Our downstairs talks were beginning to occur regularly. At first it was the normal family stuff: a recycling box I'd left in the kitchen, garbage cans I'd forgotten to bring to the curb, dishes in the sink. But interlaced with the normal family stuff lived the many unpredictable acts Dad felt we had committed against him.

He'd stopped sleeping at night, preferring his desk downstairs in the makeshift office to his bed. Nighttime, he declared, was when he worked best.

In the living room, Dad removed his eyeglasses from their case and set them on the coffee table. He was still in his work clothes. His tie was off. His shoes were on. Mom motioned for me to have a seat.

"You know why we're here?" Dad asked.

I didn't try to guess.

"Ellen, you want to tell her why we're here?"

Mom remained silent, always a good choice.

Dad kicked his feet out of his brown corduroy shoes. "What were you wearing to school today?"

I saw where this was going and looked at Mom for help.

"Mom's vest."

"Whose vest?" Dad asked.

"Mom's vest," I said, biting the skin on the inside of my cheek.

Dad clasped his hands and looked at me.

"Ellen, why was Rachel wearing *your* vest?"

Mom stared straight ahead, as if she'd been through inquisition for hours already and knew better than to speak.

"Mommy offered you her clothing?"

The old offering of clothing. It sounded like something biblically perverse.

"Yes," Mom said. "I lent her the vest."

I was settled by her confidence in sharing this detail.

"Why, Ellen? When we so clearly have a rule? You and Rachel are not to share clothing. You've got to learn your place or she's gonna walk all over you."

"I understand," Mom said. "It won't happen again."

"She is a child," he said, motioning toward me. "Don't you get it?"

"I do get it, Steve. I made a mistake and I'm sorry."

"You did more than make a mistake. You gave Rachel permission to play your part. She has way too much power in this house, way too much control over you."

"Can I leave?" I asked, aware it was a bit of a long shot.

"Can you leave?" Dad shook his head the way he had when I was young and asked if I could fly. "What do you mean, can you leave?"

"Like, can I go to bed?"

"'*Like* can I go to bed?'"

"Can I go to bed?"

"Can you speak English?"

This type of question was a conversational roadblock, a dead end on the map in Dad's mind, which both Mom and I were trying desperately to navigate.

"Dad, the clothes-borrowing thing, it's not that big of a deal."

"Not that big of a deal? Not that big of a deal that Mommy is controlled by you?"

"Mom, I need to go," I said.

"What's this *need*?" Dad asked. "What, really, in your life do you *need* to do?"

He had a habit of posing rather profound questions in small, claustrophobic moments.

"*Need* to go practice."

"Can you speak in a full sentence?"

"I need to go upstairs and practice my lines."

"Your lines?"

"For the speech competition."

Dad laughed. "Competition?"

"Yes."

"Did you hear that Ellen? Rachel's going to participate in a competition."

I'd joined the speech team, since I'd had little luck making my way into the theater department. It was a diversion from the dramatics of life at home, both because it got me out of the house for competitions every Saturday and because it got me out of myself. Acting required a cultivation and exploration of emotion, which allowed me to dig deep inside, past the anger and frustration I felt for Dad, until I hit the still and quiet part of myself I hadn't known existed.

Agnes of God was a haunting and beautiful play about a naive young nun who gives birth to a child after a Church scandal, then kills the infant, who, she's convinced, was the product of a virgin birth. Susan Payne and I took a scene toward the end of the play and entered the category of dramatic duet acting, competing with other pairs from various high schools. She played Dr. Martha Livingston, the psychiatrist who was sent to investigate the crime. I played Agnes.

I played Agnes all the time. I played Agnes against Dad, crawling into her mind and her life so I could escape my own. And through my entire sophomore year, I got to know Agnes better than any living person.

I lived through her, released her at school, where I could be myself, and then hid behind her like a shield when I went home. Acting was the closest I could come to leaving without actually running away, and the more disturbed the character I played, the luckier I felt waking up each morning in my own skin.

The cuckoo clock sounded. A blue wooden bird appeared and let out a high-pitched, irritating noise. Dad gazed at the clock, and, as if he was speaking to the bird, he said, "I bet you think you're a good actress, Rachel?"

His voice had softened.

"I do."

"Such a shame," Dad said. "Such a shame that you have traveled so much and learned so little." He slouched further into the couch, his voice no longer soft but teetering on the edge. "How lucky you have it, Rachel. With all the homeless, parentless, poverty-stricken, malnourished, enslaved children in this world."

I imagined Mom's vests draped across the bodies of the world's homeless and malnourished children. Like a car swerving out of control, this conversation was going to follow no predictable course. Dad's eyes were blazing.

"A child who has every conceivable amenity that life has to offer, and you can't follow a very simple rule."

My mind split over that. My mind was always splitting over the things Dad said. First, there was the issue of having every conceivable amenity, which we didn't, and I was thankful that we didn't. It killed me to see parents ruining their kids with TVs and phones and video games. This was something I thought Dad did right, providing us with the necessities, infusing us with values that would inform our entire lives, emphasizing travel and education. I knew it even then, even that night and in the midst of everything, that I'd forever be grateful for this. But it seemed hard for Dad to give without regretting what he had not had.

Then there was the issue of the vest, and the unfortunate reality that we would never get back to the simple subject of the vest-borrowing: an act involving an inanimate object, something that didn't require me to do much more than apologize. But unfelt apologies got me nowhere when Dad had things to get off his chest. Dad was infuriated to the point where he couldn't see either Mom or me sitting in front of him. Nights like this, we were not his wife or his daughter. We were conspirators.

"A fraud," he said, offering up the word like it was something we'd all been racking our brains for. "That's what you are. All your attempts to portray yourself as a victim of abuse. To your guidance counselor and your grandmother, to Mommy. Mommy, who's too weak to see that you're manipulating your way between us, who do you think you are?"

Mom stretched her hands above her head. "Steve, let's move on."

"It's a big, immature charade. Everything about you is a big charade."

I imagined the three of us playing a family game of charades.

"You've built a fraudulent little acting career for yourself."

Dad seemed the way God had been described in Hebrew school. A greater force, a force not human, something we could feel in everything.

"I'm ashamed to say you're my daughter, disgusted and ashamed," he said, "and Mommy, too."

On the couch beside Dad sat a small, decorative pillow embroidered in pink thread with a lace eyelet trim around the edges. It read THANK YOU FOR NOT SMOKING, and I wondered how Dad could care about something like smoking and still say the things he did. I envied the girls with alcoholic dads. So what if their dads had run their cars into trees, smashed their hands through glass, embarrassed their wives at parties. These girls could get up on their legs and walk out of their houses. I gulped an undeniable lump back in my throat, a gulp so loud it sounded microphoned. I was losing steam.

"Are you going to cry for us, Rachel? Are you putting on a little show?"

He looked like he was going to be sick. I tried not to blink. I hated crying in front of him. Jenny cried. Mom cried.

He was smiling, and when he smiled, all his teeth showed. I thought about Dad's golden fillings and my own mouthful of cavities, and how after the dentist Mom liked to remind me that I'd inherited Dad's bad genes.

The dog had curled into a ball in the corner of the room underneath the cuckoo clock.

"Here we go now," Dad said. "Look at her eyes, Ellen. Here come the tears. Now the show has begun."

His smile looked almost pained. Had he been a stranger on the street with that pained look on his face, I would have thought he was lost.

"Stop," I said, thinking about how I used to cry mercy to end a game of arm-wrestling with him.

"Stop what?" Dad asked. "Is this your winning number? Is this what you're going to perform at your little contest this weekend?"

I looked up at Mom, for a moment allowing myself to forget where I was, becoming Agnes, sweet and innocent and a little bit insane. Turning Mom into the role of Dr. Livingston, letting Dr. Livingston watch sympathetically as I lay in bed, covered in blood. Pressing my eyes shut, I concentrated on how Agnes must have felt to cradle that dead baby in her arms. I practiced my lines.

This was how I separated and protected who I was from what was happening with Dad. I excavated the emotions of characters to distract me from my own. Dad could see it, I was sure. He could feel me becoming someone else, and I think it made him sad that my reactions to him were filtered and not raw.

Standing up from the couch, he walked across the room to the thermostat. Mom and I followed with our eyes as he adjusted the temperature.

"Can I please go?" I asked.

"You're not going anywhere." He sat back down on the couch. "Not going anywhere now and you're not going anywhere in life."

Tears dribbled down my face. I didn't wipe them away and I didn't look down, and I didn't know whether or not I was crying because of the hurt or because I couldn't leave. A haunting ambiguity. I'd spend the next ten years leaving everything and everyone that made me hurt even slightly, with no explanation and without looking back.

I stared straight at Mom, silently begging her to make him quit. I no longer cared about anything other than getting out of that room.

"It's bad acting, Rachel. All this to get Mommy's attention."

I arranged my hair across my face.

Dad knocked his knees together like a little boy. "Do you like the show, Ellen? What do you think? Should Rachel become a professional actress?"

"You make me want to die," I mumbled.

"Steve, she says she wants to die," Mom said. "She just said she wants to die."

"She wants to die? You want to die? I thought this was just a show. I thought that she was acting for us."

"I don't think she's acting, Steve. She says she wants to die."

Dad stood up. "You want to die?" he asked, walking into the dining room where we could no longer see him.

"No," I said quietly.

I heard the refrigerator slam shut and the back door open

and close. Dad was letting the dog out. How, in the middle of this conversation, had he thought to let the dog out? And how did Mom, who was still seated on the couch, look so unfazed by it all, almost bored?

"Steve, are you hearing me?" Mom called into the kitchen. "Rachel wants to kill herself."

"That's not what I said," I whispered.

"That is what you said."

"What I said is Dad makes me want to die."

"That is a very serious thing to say, Rachel."

"I know," I said. "It's a very serious thing to feel. You think I like feeling this way?"

"I'm concerned," she said in a voice loud enough to reach Dad in the kitchen. "Are you really feeling like you might want to hurt yourself? Do I need to get you help? Because I will if you feel like you're a threat to yourself."

"Fine," I said, annoyed, wondering how it was we'd gotten to this point. Mom knew suicide was not my style. I would have walked out the front door before I ever considered killing myself. I would have stolen Dad's car and moved to New Mexico, or taken his money and traveled through Europe, moved to L.A. to become an actress, or opened up a crepe shop on the beach in South Florida.

"Well, then we'll have to take you to the hospital."

"Let's do it," I said, liking the thought of being anywhere other than there.

Dad returned with a glass of orange juice and sat back down on the couch. Mom, not in the voice of concern that she had exercised moments earlier but rather matter-of-fact, said, "Steve, Rachel feels that she needs help."

He drained the glass in one gulp. "If you're planning on

killing yourself, Rachel, you better hold off until you've taken care of business."

I imagined the skin of his face being eaten by worms, his eyelids falling away like dust, his lips evaporating.

"What business?" asked Mom.

Dad crossed his hands over his chest. "Tuesday night, it's garbage night."

"How's Friday work?" I asked.

"Friday's Shabbat."

"Sunday any good?"

"You've got piano Sunday. You made a commitment to Grandma."

"Hold on a minute, Stevie. This isn't stuff you play around with. Rachel's threatening to kill herself."

"No, I'm not."

"Rachel, you gonna kill yourself?" Dad asked.

I thought about the ramifications of saying yes. So what if he thought I might?

"C'mon, Rachel, is it really so bad? Is your father, who takes you all over the world, to Europe, to Africa, who sends you to summer camp, so terrible you're gonna kill yourself?"

I refrained from correcting his use of "gonna," deciding it wouldn't help matters much, but it was all I could hear. "Ya gonna . . . ya gonna?" Dad hated when I talked like that, but when he got this angry he seemed unaware of the way he spoke.

"You, a kid who has everything, you gonna kill yourself?"

I shook my head.

"Let her do it, Ellen, let her do it."

"I'm taking her to the hospital, Steve. You want to get locked up?"

"Take me," I said. All I wanted to do was lie down for a few weeks somewhere other than there.

The cuckoo clock made its noise again.

Another hour had passed. A constant landmark in our late-night living-room sessions. Dad picked a *National Geographic* off the coffee table and began to flip through it.

"Let's go, Mom. I need some fucking help," I muttered.

That was the first time I had ever sworn around my father, other than the memorable tune he'd caught on tape. And to my surprise, he seemed not to notice. He did nothing at all, just sat there and watched my face contort and crumble, like a broken machine of jumbled-up rusty parts. My lips quivered; I could feel the hiccups coming on again. An immediate release swam through my body, the calm of surrender. I wanted to go to the hospital for a couple of months, have somebody feed me, watch a little TV, piss in a pan, have someone come and take my piss away.

"I know the truth, Rachel. I know the truth."

Dad's face was so still and composed that for a moment I thought we were feeling the same deep sadness I kept isolated from the rest of my emotions.

I watched his eyes flicker, like candles burning out, and wondered what truth meant to Dad. Was it a meaning that granted him serenity or made him more afraid? Was it a thing to seek or was it something he felt he owned? To me truth was a void, an undiscovered world, choices I did not yet have the freedom to pursue, an afterlife that would eventually present itself. Truth had no past. It dangled somewhere in between where I was and where I'd end up, and that night there was only one truth that mattered: I couldn't take it anymore.

"Get out of here," Dad said. His tone was quiet enough to

reflect the hurt, yet still stern. He was a man who wanted it both ways, to be a child and a God.

"C'mon, Ellen, run to her rescue!" he said. "She needs some psychiatric help!"

He looked at me and laughed in the bitter, calculated way one laughs when something is not at all funny. He had won. We would have more nights like this. Like two wrestlers in a ring, Dad had taken me to the ground and successfully kept me down.

"Bye-bye, Rachel." He waved his fingers at me.

"Bye-bye, Dad."

"Go on, Ellen. Take your daughter to the hospital."

I was ready for the hospital, sick of this world that consisted only of us. Mom stood up and I could tell by the look on her face that she, too, had had enough.

I walked over to the banister where my jacket was hanging. Mom stuck her hand out and reached for me, though she looked directly at Dad.

"You need some help," she said to him. Then to me, "Get in the fuckin' car!"

The hospital was close to our house. Still, I was looking forward to the car ride over.

Dad paced the living room, clapped his hands together in applause. "A performance well done, Rachel."

"Let's go," Mom said. She was still watching Dad, maybe waiting for him to tell her not to go.

"You go tell the doctors about your misery and hardships, Rachel. You go put on your little play. All the abuse that you suffered at the hands of your treacherous father."

I stood on one leg in the hallway and pulled my winter boot over my foot, thinking how nice it would be to get into a robe in the psych ward and go to sleep. I drowned out Dad's words, meditated on my mounting glorious visions of hospital life: smoking cigarettes in the courtyard and fuck fuck fucking away with some half-crazy suicidal rich boy from Winnetka, whom I'd ultimately save because I was far from crazy and completely on top of my game and I'd show that rich boy from Winnetka just how spoiled and lucky and melodramatic he was being and he'd get over it and we would fall in love.

"C'mon, sick one, let's go," Mom said. "Get your gloves on."

I took my gloves from the basket on the radiator.

"You really need to go to the hospital?" she asked in that awful British-sounding voice she got when she was losing it. "Because I'm really gonna admit you."

I looked once at the dog before turning to leave. I memorized her eyes, wondered if God was in that dog, the only pair of eyes that caught everything that happened in our house. Then I stepped outside. Mom followed behind me, her keys jangling. We got into the car. She was wearing Dad's green coat. She was on the brink of becoming someone else.

The car was cold. She turned the key in the ignition. The engine choked. I didn't put my seat belt on and she didn't ask me to. In the mirror she examined her teeth. I stared straight ahead at the peeling paint of the garage door. She put the car in reverse and pulled out of the driveway way too fast, without looking behind her.

"You littah fuckin' bitch," she said. "You fuckin' littah bitch."

She turned the steering wheel to the left.

"What the hell, Mom? What are you yelling at me for?"

The front tire hit the curb.

"Stop it!" I yelled.

"Shut up, Rachel."

"Stop driving crazy, you're gonna kill me."

"You want to die anyways."

"God, Mom, I don't want to die. Are you crazy? You know I don't want to die. I just want to get the hell out of here."

"I'm taking you to the hospital, to the loony bin where you belong."

An interesting word choice for Mom; she was offended when social workers were referred to as shrinks and the mentally ill called retarded.

"Don't drive crazy!" I yelled.

"You don't like how I drive?" asked Mom, swerving across the street and screeching the car to a lurching halt. "Then get out. Get the hell out of here."

She sat with both hands on the steering wheel. Then she made a fist with her right hand and stuck that entire fist in her mouth. I opened the car door and ran down the block as fast as I could, hoping I would fall and break my face. Mom got out of the car, came flying behind me. She grabbed my coat and tackled me down to the snow-covered lawn between our house and the neighbor's. We struggled for a while, our hearts racing. Her face was in mine. "You crazy little bitch," she said, breathing recklessly into my face.

"You're crazy. You're the crazy one, Mom. Look at you. You're the one who married him."

"You want to take me down, Rachel?"

"You're already down, Mom. You're fucking on top of me."

"You think you can play around like this, Rachel?"

"What the hell were you thinking, marrying a man like Dad?"

"You don't know anything. You're just a kid. A little shit. You're a little shit!"

Her eyes filled to the brim with an almost drunken glee, the kind I'd seen in movie characters, and then she began to sing "little shit" over and over to no particular melody: "Little shit, little shit. You're a little shit." Her breath smelled sour. I began to laugh.

Mom pressed her hands into my shoulders, pinned me down so that my head lay on the snow. I squirmed from underneath her, trying to get away. She pushed her hands harder into me, her breath circling the air and her lips dampening with saliva. With her right hand she cupped her fingers loosely around my throat and laughed. But it was not the laughter of amusement. The noise was deep and pained. She leaned in closer to my face and shook her wavy brown hair with all she had until the hairs on her head were tickling my nose. I lifted up my leg and kicked her in the ass with my knee.

"You want to hurt me? That it, Rachel?"

"Yup."

"You wanna fight . . . wanna fight, little shit?"

She was so close to my face, her eyes looked terrified, specks of gray dirtied the usual blue, eyes that no longer seemed to belong to her. Laughter. She seemed right then to be devastatingly aware of her own helplessness. Letting go of me, she sat back in the snow and began to hiccup. Hiccupped and laughed.

I tried to picture how the two of us had gotten ourselves into this, rolling around on the ground like angry schoolboys. I felt sore from the cold and the pressure of Mom's body. I was ready to surrender.

"Is this funny to you, Rachel?"

"No," I said, putting my hands over my face, laughing then crying. "I don't know what it is."

"Is threatening suicide funny to you?"

"No."

"Then why are you laughing, little shit?"

"I'm not," I said.

"Yes you are. I see you."

"Mom, will you just take me somewhere? I don't want to do this anymore."

"Where ya wanna go?"

"To the hospital."

"You're not going to the hospital."

"Well, can you at least move off me then? I can't breathe."

"You don't need to go to the hospital."

"You're right," I said, "Dad should go to the fucking hospital."

"Dad should go to the fuckin' hospital," she said. "Ya both should go to the fuckin' hospital."

I hated hearing her swear.

"Then take me, Mom. Take me somewhere!"

"You're fine, Rachel. You're just fine. You're gonna be just fine."

"Stop talking in that British voice, would you?"

"You think you got it bad? You think you got it so bad?"

"Yeah, Mom, I think it's pretty bad, and I think he's getting worse."

"Well you know what, Rachel?"

"What?"

"I'm not him. I'm me. I'm Mommy. And I can only do the best that I can do."

"Well, then your best sucks," I said calmly, my body deflating a bit.

"I'm sorry you think that."

"Look at us, Mom. You're sitting on top of me in the fucking street. I'm soaking wet and this is crazy."

"You're the one that wants to go to the hospital 'cause you think it's so bad. You think it's that bad. You really think it's that bad? You think I'm so bad?"

She didn't want me to answer. I think she also needed a break from the house, because there she was on my chest, with her eyes half-open, and sad to the bone. Tears spilled down her colorless cheeks.

She pushed her hands into the ground and stood up. Her poor white hands, she shouldn't have been anywhere close to cold like this after the frostbite. She'd gotten it when she was younger. Even in the swimming pool her hands turned white, even washing dishes when the water was too cold.

"Look at your hands, Ellen! Why don't you put some gloves on," I yelled.

"You don't care about my hands, Rachel. You don't care about my hands."

I pulled myself up to meet her. I wanted her to put me in the back of the car and drive around for a while with the radio on, just drive me the hell out of there.

"Why do you let him?" I whispered.

"Go inside," she said.

"C'mon, Mom, take me to Uncle Arthur's. Take me somewhere else."

"It's time to go inside now," she said.

"Can we just go for a drive? Just for fifteen minutes so I can clear my head?"

She was calmer now. She looked like she had risen from a long noon nap. She turned away, walked over to the car, and took the keys out of the ignition. She tried to put her arm around me but I ducked underneath, walking slowly behind her back into the house. Dad was in his office. I walked upstairs. In the bathroom I removed the wet clothing and wrapped myself in a towel. For a long time I stood brushing my teeth, letting the blood from my gums pool at the bottom of the sink.

EIGHT

The next morning Mrs. Pollock called me out of my math class and into her office. The sun trailed through the slits of her venetian blinds. She slid her glasses back on her face, pushed her chair away from the desk, and said, "I received a phone call from your . . . father."

I liked the way she paused when she said "father," as if the term itself was questionable.

Dad had called Mrs. Pollock to tell her I'd threatened suicide.

"He seemed concerned and obviously I am too. Are you thinking about suicide?" She removed her glasses and placed them in her lap.

I wondered if Dad had placed the call from home or from work.

"No," I said.

"Even in a very desperate moment?"

I loved how adults put words like "very" and "desperate" together. It sounded like a Hallmark card for a suicide occasion. "No," I said. "I'd definitely rob a bank and take off to Europe before I thought to kill myself."

"So you think about leaving?"

"All the time," I admitted. "But I won't."

I pictured Dad stewing around the hospital with his stethoscope dangling over his chest, contemplating my suicide plans.

"I promise you, he knows I didn't mean what I said last night. He hates that I talk to you. He thinks I tell you lies and he wants you to know his side of the story."

"Let's put your dad aside," she said.

I imagined Dad's body being lifted from the table on a long, rectangular cheese platter. "Let's forget your dad," she said. "This is not about your dad."

"But it is," I said. "It really is, because I'm fine, Mrs. Pollock. Inside, you know. I really am."

I liked to picture Mrs. Pollock testifying for me at court as I emancipated myself or taking me into her house for a couple of years while I finished up high school, though it was probably against some code of ethics, plus she had kids of her own.

"Let's talk about you," she said. "Are things getting worse?"

I tilted back in the chair. "My mom's so out of it, she's like another child, and I hate being near him. I really do. But I'd never do that, I just wouldn't. I like life."

"Do you talk to anyone besides me?" she asked. "About what's happening at home?"

"A couple people."

Uncle Arthur had tried to get us into counseling. Mom would call Arthur and he would drive to our house, almost always with Cousin Debby in tow. Arthur was a social worker and knew several professionals he could recommend. He said he could introduce us to someone who specialized in parent/child relations.

"I thought it was normal stuff," Uncle Arthur told me fifteen years later. "Rebellion, typical teenage behavior stuff, you know what I mean."

There was no way Arthur could have known. I never spoke up. Not to him and not to my cousins.

But Arthur told me later that Mom called him on several occasions to come break up their fights, to try and convince Dad to get counseling, all before I was even five. Jill and Debbie remember being in the car with Uncle Arthur driving, and Mom up front crying and swearing.

Twice, over the course of several months that fall, Mom had taken me to psychiatrists at Michael Reese HMO and both times those psychiatrists had advised Mom to get me out of the house and away from Dad until he agreed to get help. Both times, we walked out of their offices and never returned. The last visit, Mom dropped me off at Uncle Arthur's.

Aunt Jo Ann opened the door. I don't remember what was said, just that I got out a textbook and pretended to read while Jo Ann and Mom spoke outside. Then Mom came into the living room, kissed me good-bye, and left. Debbie was home. We said nothing to one another. She did her homework at the kitchen table. She spoke on the phone to her friends, and I thought how lucky she was to be able to move that freely.

If I lived with Art and Jo Ann, I would join the cross-country team, keep on volunteering at the hospital. And I knew that for things like this, with people like Art and Jo Ann, I would be rewarded. And I wanted to be rewarded, because for what it was worth, I thought I was good and desperately wanted to be seen as good.

Dad rang the bell around ten o'clock. Aunt Jo Ann told me to get under the bed. And though I don't remember this, Uncle

Arthur said he protested, because he was scared of the legal ramifications—that they would be accused of kidnapping, and he didn't want to be held responsible.

And then I was under the bed, holding my breath, as Dad walked in and out of every room. I could hear Dad's feet moving through the hallway, as if we were playing hide-and-go-seek, as if at the end of the game I would be tickled or chased or thrown over somebody's shoulder.

Dad was standing very close and I could no longer take myself seriously under that bed. I stuck my arm out, scooted my body out from underneath, rolled onto my side, and pushed myself up.

"C'mon," Dad said. "Let's go home."

I don't remember him being mad. I only remember walking through the living room and packing up my books and saying nothing to anyone as I followed him outside. In the car he turned on NPR. He asked if I had homework. He said they saved me dinner.

Mrs. Pollock considered the stain on her shirt, brushed it with her wrist.

"I'm concerned about you," she said.

"I feel stuck," I said, "stuck as hell."

I wanted to smack my head against the particleboard wall in Mrs. Pollock's office until I knocked myself unconscious. Then she'd have to drive me to the hospital, drop me at the psych ward with the courtyard and the wheelchairs and the cigarette smoke.

"Stuck as hell," she repeated, still fingering the stain on her shirt. "Well, stuck as hell is not a good place to be."

As I was sitting with Mrs. Pollock, my anger wrestled with sadness; it washed through my belly, coated the inside of my throat. Dad's phone call was like a declaration of war; a strategic way for him to invade a small region of a world I owned completely. I'd been talking with Mrs. Pollock regularly since the phone-swearing incident. She was warm and patient and I trusted her immediately. But Dad's phone calls made me realize that he'd consistently find ways to enter the areas of my life I sought to keep him out of, like a raccoon getting into the trash.

We sat there for a good fifteen minutes, saying nothing. She handed me a box of Kleenex, and she stayed with me, saying nothing, while I tried and failed to keep myself together.

"Would you rather be somewhere else?" she said after a while. "You have cousins that you're close to, right?"

"He won't let me go."

Mrs. Pollock pursed her lips together. I loved her then. For being the only person who seemed to understand how wrong things were.

"It doesn't matter what I do," I said. "There's something in me he just doesn't like. It's not getting better."

"It's not getting better," she repeated.

"He's not getting better," I said.

"I know," she said, nodding at the floor. And I knew she would help me. "I don't think it's you."

NINE

Mimi was the social worker who came to pick me up. I don't remember how the four of us came to be standing in the living room. I just remember Mimi telling me to go upstairs and pack myself some clothes and that she needed a little time alone to talk to my folks. That's what she kept calling them, *my folks.* I could hear their voices as I moved around my bedroom, grabbing underwear and socks, assigning sentimental value to items I'd never noticed before, like a black-and-white photograph of my great-grandpa and a bag full of foreign coins, which I packed in between sweaters.

Downstairs, Mimi stood in her heavy leather jacket, dangling her keys.

"You all set?" she asked. "You're gonna come with me."

It was quite a declaration. I didn't know where we were going. I didn't care.

She was short and black, with fiery dark eyes. She commanded immediate respect. She seemed unfazed by Dad. It was a fearlessness I had only seen in strangers, like the two old women from Florida who sat next to us at a dinner show in Cancún. Dad had asked them to put out their cigarettes.

They refused. An argument followed. The women took turns yelling at him about their constitutional rights while blowing a constant stream of smoke in Dad's face. Dad let out an exaggerated cough. Jenny and I were in hysterics, more amused than embarrassed, until the women started making fun of Dad. They'd turned their attention from him to each other, lit up two more cigarettes, and begun to laugh. In that moment it was very clear that Dad had lost the battle. He sat there in defeat, strangled by smoke. All of us did. Then he picked up his camera and started taking close-ups of the women, shooting photo after photo until he got them roused again, at which point they each stuck their middle fingers up to his camera. I felt something in between delight and deep regret. The women were patently unafraid of the most powerful man in our world.

"You ready?" Mimi asked. I tossed the duffel over my shoulder.

I have no memory of Jenny that night. I have little memory of Jenny at all during those years. I know that I envied her good fortune in not being Dad's chosen one. I know I envied her ability to hide.

Dad stood with his arms crossed and his sweatpants sagging, examining our living-room furniture, as if there was something curious about each piece. Then he turned his attention to his New Balance gym shoes, the broken sole and the body of the shoe held together by two thick pieces of duct tape. I looked at the hole in the ceiling, the damaged plaster still dangling from above, where a family of raccoons had camped out. They'd made their way into the garage and up the walls before the ceiling caved.

I wondered at what point Mrs. Pollock called Mimi. Was it immediately after I'd left her office? Or was it after she ate her lunch? Had she discussed it with someone else or made the decision herself? Had Mom and Dad had any warning?

"All righty then," Dad said. "All righty then."

Dad never knew what to say. "All righty then" was what he said when we left someone's home after dinner. We'd be stuffed and exhausted, standing by the door, buttoning up our coats, kissing each other good-bye, talking about how nice that was, how we should really do this more, and when it came Dad's turn to kiss someone or speak those words before good-bye, he'd say one of two things: "All righty then," or "Take a couple aspirin, call me in the morning." Then he'd stare at his shoes and stammer about until our hosts shut the door behind us. Alone again in the quiet of night, just the four of us walking back to the car, Dad would let out an exhaling gasp and say, "Phew."

I followed Mimi out the door.

Her car was small and brown and smelled like soap. The seats were off-white vinyl, the color of dirty teeth. The steering wheel was dressed in a cheap fuzzy cover. As we backed out of the driveway, Mimi put her arm around my seat and turned her head to look for oncoming traffic.

"How you doing?" she asked, placing her hand on my shoulder, giving the shoulder a little squeeze.

I felt myself relax.

"Have you eaten?" she asked.

"No."

"You like Greek food?"

"Yes."

"You want to stop at Crossroads?"

"Yeah," I said. "Good vinegar fries."

"You want to grab a bite in the café or take it in the car?"

"Car," I said.

She was relieved by this answer, I could tell.

"You want to eat first and then we can talk?"

"We can talk," I said.

"Mrs. Pollock and I did some talking earlier on today," she said. "And she felt that you could use a little break from home right now."

"Right."

"Do you know what DCFS stands for?" she asked.

"Department of Children and Family Services."

"And do you know what DCFS does?"

"Yeah. My mom's a social worker. Hard to believe."

"Oh, I believe it," Mimi laughed. "We're some of the most screwed up of the bunch."

She pulled into a parking space just outside Crossroads, picked her purse up off the floor, and asked if I wanted to come in. I stayed in the car, wondering if she was going to pay for dinner or if she'd been given vouchers for this type of thing.

She left the keys in the car so I could play with the radio. I remember thinking that she liked me or at least considered me an easy case, trusting me enough to leave those keys. The night was lit by snow and the yellowish tinge of dulled-out street lamps. I was terrifically aware that I was going somewhere. There was a woman standing in a café, who was going to return with food and then take me to a place that was anywhere other than home. I felt still and revived and emptied of all that I knew.

Mimi returned drinking a can of Orange Crush, holding a grease-stained bag of vinegar steak fries and a spinach pie. "Dig in," she said.

Something like excitement dusted the insides of my bones. I

felt like I had just won the final round of a game show and was being whisked away on an already-paid-for vacation. I offered her money for the spinach pie, which she declined by swatting her hand at me, saying, "Please, girl."

We drove west, away from Lake Michigan, away from the high school and out of Evanston, through Skokie, past Lincolnwood, into Niles and then through Morton Grove. We passed the pink mini-golf joint near the Pizza Hut where we often ate. We passed the Dairy Queen, followed by an abyss of drab-looking condominiums and townhouses.

"The place we're going's called The Harbor," Mimi said. "It's transitional. You'll be there for a week or two and then we'll figure something else out. It's a good place, really laid-back. Only nine girls there right now. Some places are so crowded it's not even right."

"What's it called again?" I asked.

"The Harbor."

At least it's not the Blah blah blah of Blah blah blah for fucked-up girls, I thought.

"It's a cool place. Like I said, it's really laid-back. You doing all right?"

I didn't know if I was supposed to answer.

"Hey?" she asked. "You okay?"

"Yes," I said and meant it.

"Good. What station do you want?"

What station do you want? It was the question I would have asked the kid I'd just removed. I considered the options, relieved that I didn't have to do all the talking. That was her job. I could just look out the window and listen to the radio.

"Magic 104?" This was my compromise station with Mom, all oldies all the time.

"Seriously?" she asked. "All you guys usually want WGCI or B96."

I shrugged, touched that she had clumped me in with "all you guys."

"Go ahead, turn it on," she said.

I got grease on the radio knob, then pulled out a napkin to wipe it off. "Let it be," she said. She probably pulled this type of ease off every night of the week with every kid she met. Impressive, I thought, wondering what the inside of her Evanston apartment looked like.

Radio commercials sold us ladies' night at clubs out in Montrose, water beds, cars, and carpets. We passed Main West Township, where last year's regional speech-team competition was held. Maybe I would go to Main West for a while, be the new kid who was automatically popular just for being new. I wasn't popular at Evanston. I wasn't much of anything.

"You know what my job is?" she asked.

"Social work."

"But you know what my job is while you're here?"

"Not exactly."

"I'm your contact. If there's a problem, you get on the phone and call me. If you have a problem with the girls or the staff, if you need something, like toothpaste or socks, if you just want to call me to talk."

"Thanks."

"That's what I'm here for."

I was not concerned with much of anything aside from the impression I was making on Mimi. I didn't want her to feel sorry for me, but I certainly didn't want her not to.

"So what's Operation Youth Umbrella?" I asked. "I saw it on your car."

"It's an outreach program for kids who need some guidance, some activity."

"Like an after-school program?"

"Yeah, we've got tutoring and games and stuff like that."

"I volunteer at Cabrini Green each week," I said.

"Really? What do you do with kids at Cabrini?"

"Art projects usually, things to keep them busy until their parents pick them up."

"You know what I think?" she said. "I think you'd be good at my job."

"Really?" I sat back in my seat.

"You seem to have a lot of patience."

We were quiet for the rest of the ride. I no longer recognized where we were, just that we were very far west and the suburbs were long and flat and full of identical houses.

There was no sign in front, nothing that made The Harbor any different from the other houses on the quiet, tree-lined street. I walked in back of Mimi, wondering if the neighbors told their dinner guests they lived next door to a girls' home. Or did they confuse it in that generous, well-intentioned way, and call it a shelter? Correctional facility? Rehab? Did they say things like "They're very nice girls for being where they are."

A paper turkey ornament with moving limbs was taped to the door, which was answered by a fat woman in a black macramé sweater. She extended her hand and introduced herself, and then brought us into the kitchen, where another woman sat. The other woman was tanned, with hair down to her butt and a thick foreign accent.

A nervous excitement filled me, similar to what I'd felt on the first day of summer camp. But summer camp belonged to a different world, and looking down at the kitchen table littered with Popsicle sticks and glitter, I was well aware that this place was not of that world. I remembered the class field trip we took to Cook County Prison, when we walked around the cells looking at the prisoners as if we were at the zoo. How my visitor's pass had burned into my skin.

Fat Macramé smiled in a way that reminded me of the soccer moms who volunteered to do our routine lice checks at school. I could tell that she was something like the head of that house.

"We made boxes today," she said, indicating the mess on the table.

"Like for jewelry?" asked Mimi, raising her eyebrows.

"For anything. The girls wanted to make Christmas gifts for their families."

"How cool," Mimi said, looking at me. "Rachel teaches art."

"We've got lots more supplies," said Fat Macramé.

Mimi and I were given forms to sign and I sat at the kitchen table, signing without reading.

"You know you can call me for anything?" Mimi said, leaning in toward me.

"There is something I have to do this weekend," I said, "something school-related."

"Is it at school?"

"No, it's somewhere kind of far, I think."

"Can you tell me what it is?" she asked.

"Speech-team competition," I said, "state championship."

"We'll make that work," she said.

I thanked her for picking me up, in that same awkward way I'd offered her money for the spinach pie. She seemed humored by this and waved me off.

■

Fat Macramé showed me around the house. We walked through the living room, past a few girls watching TV, and up to a bedroom. The room was small, with four beds, three of which looked like traveling circuses, filled with the beady-eyed stuffed animals that boys won for their girlfriends at carnivals. On the wall there was a poster of Bon Jovi in a bubble bath. On the dresser was a mess of lipsticks, spray bottles of perfume, and nail polish. Near my bed sat a silver safe. Fat Macramé went down the hall and brought me back some sheets and a lock. "Sometimes people take things," she said, which reminded me of a book Mom had bought for me when I was a kid, called *Sometimes I Get Angry*.

Alone in the room, I covered the mattress with soft, worn sheets and locked my journal in the safe. In the bathroom I changed into flannel bottoms and a sweatshirt. For a long time I sat on the toilet and tried my best to cry, but I was not sad, and anyway, I did not cry when I was sad. I cried when I was stuck. I cried when I felt wronged. I cried when I no longer had it in me to fight back.

Downstairs I took a seat on a footrest next to Fat Macramé. It was ten o'clock and they were watching *Night Court*, which I'd never seen, and I couldn't help the satisfaction I got from missing *Nightline*.

I'd walked into the middle of a conversation, which looked more like a fight, but soon I'd learn that sarcasm was the law of the land at The Harbor; everything was said with a confrontational edge even if there was no confrontation.

"I got a name," said Fat Macramé to one of the girls on the floor. "I know you don't know that, but I got a name."

The girls were cracking up and Fat Macramé was trying her best not to laugh.

"And if you want me to turn the volume up, you can ask me by my name."

"I know you have a name, it just don't always come to me on the spot," said one girl. "You know what I'm saying?"

"What's my name?" asked Fat Macramé. "What's my name then?"

"I know your name," said the girl.

Looking at the girls and gesturing to me, Fat Macramé said, "Does anyone care who this is?"

After a couple of seconds she leaned in toward me and said, "I guess they're just shy."

"Please, we're not shy," said a big black girl, getting up to shake my hand. "What's your name?"

"Rachel."

"Rachel. Pleased to meet you."

I looked back and forth between the black girl and Fat Macramé.

"Now you tell her your name," said Fat Macramé. "That's how you make a friend."

I sat in a brown recliner while Fat Macramé went over the rules. I could use the pay phone whenever I wanted unless someone else was waiting. I could watch TV as long as the others in the living room agreed to the same show. I could stay up until eleven. I could eat whatever and whenever I wanted. I could smoke cigarettes in the kitchen.

I could not have visitors without permission. I could not have male visitors on any occasion. I could not smoke in the living room. I could not light fires. I could not do drugs. I could not leave unescorted. I could not sleep anywhere other than The Harbor. I could not run away.

I was introduced to Kimmie, one of my roommates and the only other white girl. Her mouth was full of tiny rectangular teeth and there was something about those tiny rectangular teeth and the way she repressed her smile when she looked my way that made me want to know her.

"Who else sleeps in this room?" I asked Kimmie before the lights went off.

"There's an Asian girl, but she don't talk to us. I think she's on the phone right now. Oh and there's another girl, that girl's in trouble. She took off last weekend. She's got some boyfriend she's not supposed to see."

"Are they looking for her?" I asked.

"I don't know. I guess. They're real pissed. She's fucked up."

"How?"

"Mean. She's real mean. She's the only real mean person here. I'm telling you, you don't want to cross that bitch."

Kimmie's hair was so blond it looked white, and I watched her run her fingers up her legs before I became aware that I was watching. Then I stopped watching. Her hair fell over her tomboy knees and I knew there was something hard inside her that would always be hard and I wondered about the things she'd seen.

"You got a boyfriend?" she asked.

"Kind of," I lied.

I'd just had my first real kiss.

John had a reputation. He cheated. He stole. He broke into cars and houses and other guys' girlfriends. I never was sure why he took an interest in me; we'd never even talked. But one day after school, dead and dark by four p.m., I saw John seeing me. He was standing with his skateboard. I was waiting for the bus, for warmth, for anything or anyone who might pick me up and deliver me elsewhere. "Who are you?" he asked. He was so tall I had to bend my head back just to answer him. He put his hands on my shoulders and repeated my name, and I could tell, by the way he bit his bottom lip and smirked, by the way his eyes glassed over my face, that there was something he had that could make me break inside. After that day at the bus stop, he bought me a white linen blouse from the Gap on Howard Street and asked me to his house. My friends said the blouse was shoplifted but I didn't care. I'd go to John's every day after school and we'd put on Led Zeppelin tapes. He'd lie on his bed like a king, with his shirt off, and his hands behind his head. He liked it when I wore the blouse. Said it made him feel like I was his. We never had much time, because of my five-thirty curfew; still, I enjoyed the thrill of closing my eyes and becoming entangled in another human being, how dark it was underneath my eyelids and how little of myself I remembered in his arms. In December John announced that I was sweet but he was bored. We were moving too slowly. Black Dog was playing and John was batting around at my bra, which I'd draped on his low-hanging ceiling lamp. "Put your hands down my pants," I said, and with his fingers deep inside, I placed his pillow over my face and cried. "Feels good," I said when he asked what I was doing. "Really really good." But all I could feel was the hour caving in, collapsing, then the cold on my face as I waited for the bus to bring me home, and the inevitable sensation of sinking when I put my key in the door and smelled dinner.

"What's wrong?" Kimmie asked. "You don't go with him anymore?"

"No, I do. You have a boyfriend?"

"Hell no," she said. "I don't want no baby. Are you pregnant?"

"No."

I hadn't given sex much thought. When I did, it was in the moments I most wanted to disappear, an escape from the late-night living-room conversations. I didn't associate it with pleasure or pain, but a chaotic, almost violent release, something that would knock the thoughts right out of my head, something heavy weighing upon me. Aside from one silent lanky boy in my study hall, I'd imagined it with no one in particular.

"Seems everyone but me is pregnant," Kimmie said.

I thought to tell her I'd never come close but I liked giving her the impression that I had.

I lay on a mattress that was not uncomfortable, in a room that was dark except for a spot on the wall lit by a nightlight, something I hadn't seen since I was little.

I thought of a trip my family had taken to the Grand Canyon when I was ten. How it was hot and crammed with people, and littered with tour buses, mules, and T-shirt vendors.

"Can you believe these people?" Dad had asked, pointing at the people on the mules. He was right. They were a fat and content-looking herd. "I really can't believe how lazy they are," I said, "or their whiny kids." Loud, obnoxious voices boomed from mouths stuffed with hot dogs. Dad and I were repulsed. "Is this what they came to do at the Canyon?" I asked Dad, as we walked briskly in front of Mom and Jenny. I could tell Dad was very proud of me. He hated bratty kids. He hated smokers. More than anything he hated complainers.

It was after we ate our peanut butter sandwiches and were about three-fourths of the way down to the bottom of the Canyon that Mom started in with how cold it was getting and how she didn't want to be stuck in the Grand Canyon when it got dark. It was very dangerous, especially on such steep cliffs.

And it would take us even longer on the way back up. "It's just not smart, Stevie."

"Why are you discouraging them, Ellen?" he asked, lowering his voice to a loud whisper. "That's the last thing the kids need . . . their mother telling them that they can't do it."

"Steve, it's not that I think they can't do it. I just don't know."

"You just don't know, Ellen. You really don't. You really, really don't. We've come all the way here to hike the Grand Canyon and you want to turn back?"

I understood Dad's frustration. There she was, discouraging us from pushing our limits, doubting Dad's ability to lead us through the Canyon after dark. We walked for another half an hour.

"Stevie, this is getting ridiculous," Mom said. "Look how dark it's getting."

"Is it the darkness or are you just tired?" Dad asked.

"Both," she said.

"You want to take a mule back to the top, Ellen? Or should we get an emergency helicopter for you two? I mean, there's not much of an option here, and Rachel and I are going to the bottom."

Dad had started to run, kicking dust into clouds, racing toward the bottom of the Canyon. Then Mom began to run, chasing Dad as he ran faster, chasing the fat pink sun as it slipped into evening. I ran after Dad, my arms flailing everywhere, chasing after the respect I'd earn in not being a quitter.

When I woke that first morning at The Harbor, the weirdness of it hit me, like the heat of warm water against cold hands. Kimmie and the Asian girl were still asleep. It's not that I didn't know where I was, it's that I didn't know what to do next. I didn't know who I was supposed to be, but it wasn't the girl at the Grand Canyon. That wasn't going to earn me respect at The Harbor.

I walked down to the kitchen, where the longhaired woman sat at the table, reading the *Chicago Sun-Times*. There was a difference between people who read the *Sun-Times* and those who read the *Tribune*. Most people we knew read the *Tribune*. Dad read the *Tribune*.

"Breakfast's help-yourself," she said, stamping out a cigarette in the dead, flaky bottom of an ashtray.

Economy-size boxes of Pop-Tarts, English muffins, and Cheez-Its lined the counter like miniature army battalions. I went in for the Pop-Tarts, thinking, as I placed them in the oven, about the role that food played in defining what sector of the world we belonged to. I'd put money down that the food in Kimmie's mother's refrigerator was not the food Mom fed to us.

Long Hair lit another cigarette.

"The girls start class at ten."

"Here?" I asked.

"Close by," she said. "Just for the girls."

The thought of it made me want to crawl out of my skin. I tried to picture sitting around and learning math the same way we'd sat around and watched *Night Court*.

"I can't do that," I said. "I've got things at Evanston."

"We've had girls commute," she said. "But we've got to get permission from your school counselor."

That morning I watched *Sale of the Century*, *The Price Is Right*, *All My Children*, *Days of Our Lives*, and *Donahue*. Then I had a sandwich and returned for *Santa Barbara*, *Love Connection*, *Family Feud*, and *People's Court*.

■

The public bus to Evanston High School took an hour and a half each way. It was a ride I loved. Time that seemed to be still, complete, and full of hope, both then and in how I'd remember it years later. The bus passed through the suburbs, into other people's mornings. I watched men wiping snow from their plastic-wrapped newspapers and children crowding sidewalks in their Catholic school uniforms. It amazed me how much living we had to do in one lifetime; the idea that I could be anything other than a child was foreign to me. The idea of being a mother, a wife, seemed almost distasteful.

I was lonely. I felt it deeply and permanently, that this state of being on my own might never disappear. But I welcomed the loneliness, which had everything to do with the comfort I found in being anonymous. There was room for it on the back of that bus. There was room for it at The Harbor. It's never loneliness that nibbles away at a person's insides, but not having room inside themselves to be comfortably alone.

There were hours of television each night. I sat in a blue chair, foam coming out of the arm, and watched the girls watching TV, laughing and hollering and talking back to the thing. I watched shows that didn't interest me, just to hear the girls recycle jokes from reruns, making everything a little funnier than it was before. I watched who laughed at what, who laughed habitually, who laughed quietly from inside, and who couldn't find it in her to laugh.

It snowed more than usual that year. The bus driver drove cautiously. I'd brought one tape with me to The Harbor. It was Cat Stevens, and I listened to the same song over and over again on that long, peaceful ride to school.

Oh baby baby it's a wild world,
it's hard to get by just upon a smile

I wondered what it was Mom needed to get by. I had friends whose mothers supported two kids on a single teaching salary. Dad was a doctor, he could pay child support, he could keep the house if that was what he wanted, and we could see him once a week on Saturdays. I never knew how Mom and Dad worked out their finances, if they had separate bank accounts or if he monitored her money like a child's allowance. She said she had a spending problem, and that shopping was an outlet for her sadness. It was true she bought a lot of stuff. Hair products she'd use once and throw away, sweaters she wouldn't take the time to try on, and gifts I didn't ask for. I wanted her to save money so we could leave Dad. I returned the things she bought me for Hanukkah and birthdays, I returned the things she bought for herself and never used, like bottles of Paul Mitchell hair gel. I saved the money I made babysitting, and then I'd stuff it in her purse and in the pockets of her jeans. I'd leave $20 bills crumpled in the laundry pile, so it looked accidental, so it looked like it belonged to her. I'd been doing it for years, all the time thinking that if she just had the money she would leave.

I knew what I needed to get by. College was the prize I was holding out for and I didn't want to screw up the best thing Dad could give me. Mom was always reminding me how soon I'd get away, and how expensive my getaway would be if I had to do it without Dad paying my college tuition. To drive the point home, Mom said Dad owed me. That I should at least stick around to get the thing I deserved. "After all this baloney," she would say, and hearing her put it that way made sense. I had access to things the girls at The Harbor didn't. I'd been promised a better return for my investment.

■

I was given permission to stay late at school to practice for the speech competition, so by the time I got back to The Harbor each night it was already dark. I'd make a couple of sandwiches and eat them at the kitchen table. Kimmie sat beside me, smoking cigarette after cigarette, while I finished my homework. One day she looked down at my geometry book and said, "You must be old, that shit looks hard," which was how I learned she was only thirteen. It hadn't occurred to me she had an age.

At night I emptied quarters into the pay phone and called friends. "How come you get to talk so much tonight?" they asked. I told them Dad was out of town.

Like Mom, I did not want to ruin other people's notions of who we were. All it took was imagining Mom's awkwardness in having to explain my absence. But Mom had already laid the groundwork. She was a strategist at covering up what was going on. I'd learn it later, from family friends. I was "troubled," "out of control," "rebellious." These were things all parents could empathize with, behavior no one would question, conditions which would explain my absence for the next fifteen years, well into my adult life. I was shocked to find this out, buried in an anger I couldn't confront. By that time I knew there was no clarity to be gained in discussing things like this with Mom. She'd always find a way to justify her decisions. It was how she survived. Still I was amazed that she didn't confide in her close group of friends. Not for my sake but to relieve herself of some of the burden.

I babysat for the son of a woman named Susan who lived in the house next door. Mom and Susan had been taking morning walks for years and I imagined that Mom was confiding in this woman, who'd been divorced and remarried, and that perhaps she was advising Mom to do the same. But I saw how Susan smiled and waved at the four of us, with something close to adoration in her eyes, and I could tell she knew nothing about us.

I imagined Susan coming to our door with a babysitting job for me, and Mom explaining I'd moved out and offering her Jenny instead. Then I pictured Susan asking where I'd moved, to which Mom, planted in the hallway with both hands on her hips, would say, "She's been adopted."

I liked the idea of adoption. Being placed with a veal-eating, tennis-playing, churchgoing family, getting set up in their guest room, called to dinner at their table. Would they expect me to celebrate Christmas or would I go home to Mom and Dad on Passover? Would I still be welcome? Would we come to an understanding that my leaving had been for the best? If so, I was willing to take the blame. For being rebellious, troubled, whatever it was that sent me to the veal-eating family. Or would I be matched with Jews? Jews liked to adopt. It was a ridiculous thought at the age of fifteen, so soon to be on my own, but I wanted to be loved as badly as—if not more than—I wanted to escape. And wanting something brought on more despair than wanting nothing.

I had friends whose stepfathers treated them like their own children. I knew I could be loved in that same way, and I liked to think that whatever random family ended up with me would feel some sense of relief that they got the kid without the issues, the kid who volunteered at the hospital, got cast in plays, didn't crash their car or smoke or go to bed with boys. I liked to think that random family would become so attached to me that by the time I was walking off that stage, diploma in hand, out of their home and on my way to college, I would be their child. That's what I wanted, to be somebody's child.

■

On the Saturday of the State Speech Team Finals, Mom arranged to pick me up from The Harbor and bring me to the meet. In the car I changed into the outfit she'd brought me from home and went over my lines in my head. "I guess you don't feel like talking," Mom said.

I shook my head. It hurt me to see her. To see that she could pick me up and drop me off, but she could never save me.

Susan and I were competing with the scene from *Agnes of God* that took place in a convent room after Agnes wrapped the umbilical cord around her baby's neck and killed him. Because Agnes is in such a terrible state of denial and confusion over the pregnancy, she confesses it all to the nun in a trancelike state, and the audience comes to learn that Agnes doesn't even realize she's killed her own baby.

The state competition was more serious than the regional. Unlike the usual three judges, there was an entire panel scoring our performance. We were competing against twelve other pairs for dramatic duet, which meant a trip to Florida for the nationals if we took first place. My goal was to get deep enough into Agnes's mind that I'd cry real tears toward the end of the scene. I wanted to cry. I had been trying all season and I didn't think we had a chance of winning unless I could hit that point of emotion where believable tears came.

Our coach Mrs. Holden told us never to force it, that tears would come if they were real and there was nothing worse than faking it. Mrs. Holden watched us practice. She'd stop us in the middle when she thought we were not focused on the thoughts behind the words. "Use your real-life emotions to develop this character," she'd say. "That is all you have to work with."

Susan and I performed first, and I could hear nothing but my own voice reciting memorized lines. I could hear the girl in the next room warming up for the prose competition. I could hear the steady tapping of a competitor's foot. I could hear Susan's voice questioning me about killing the baby and she sounded like herself and not the nun. I could hear everyone and everything but Agnes.

I imagined the judges looking down at their score sheets and taking notes on just how much I was fucking up our performance.

Toward the end of the scene something broke inside me. It was when Agnes falls to the ground and recreates herself giving birth. I dropped down to the floor and curled up in the fetal position. Instead of finding a spot on the wall to stare at while I recited my monologue, I closed my eyes completely. I forgot about my lines and I forgot about myself, and in the blackness of it all, I saw this tiny baby in my arms, this baby that I'd pushed out of my body, and it had no face, no eyes, and no lips, just skin. And with my eyes shut, in front of all the judges and the other actors, I slowly let myself go.

I didn't worry about going overtime, one of the factors for disqualification; I stayed on that floor until I fully became Agnes, until I seemed to start to cry. When I opened my eyes, Susan, who was no longer Susan but the nun, had me cradled in her arms. Staring down at the floor, I paused until the words filled my mouth. Then I said my last line. We lowered our heads, to the noise of clapping hands. We looked up at the judges. I was sure that we had won.

We took twelfth place. Almost every judge wrote that Agnes didn't come alive until the end of the scene. We were each given a plastic trophy, which I gave to Mom on our trip back to The Harbor.

"I'll trade you," she said, reaching into her blue bag and taking out a menorah. "This is so you can celebrate Hanukkah while you're away."

I put the menorah in my lap and turned on the radio.

"Is this your way of telling me that you don't want to talk?" she asked.

"Yes."

"Well, that makes me sad, Rachel."

"Good," I said. "We can be sad together."

"I have some more goodies for you even if you'd rather listen to the radio than talk to me," she said, handing me a box of wax candles and a bag of chocolate coins.

"I have a tape I want to hear."

"Okay," Mom said. "Who is it?"

"Madonna," I said.

I had it rewound to the exact place. I pressed my back into the seat, closed my eyes, and belted out the words.

Maybe someday when I look I'll be able to say, you didn't mean to be cruel. Somebody hurt you too.

Mom waited for me to open my eyes. Then she touched her sleeve to my face and wiped off my tears.

"I think that you're trying to tell me something," Mom said, smiling. "When you were young you used to draw pictures of people without arms and legs. And when I asked you what about the arms and legs, you'd say you just forgot them."

"Well what do you think that meant, Mom? You're the art therapist."

"I'm not sure what it meant," she said. "Arms and legs are typically a sign for mobility."

It always surprised me that Mom brought conversations like this upon herself.

"I've got a dreidel somewhere in the bag if you want to take a look," Mom said.

"Who do you think I might play dreidel with, Mom?" I turned up the radio, intent on feeling sorry for myself.

"This is obnoxious, Rachel. This is really obnoxious. I've planned my whole day around driving you back and forth."

"Driving me back and forth from The Harbor," I yelled. "From the girls' home."

"There's no need to yell."

"Yes, Mom, there is. There is a need to yell. Don't you get it? Things are not all right. Dad is not okay and I am in this fucking girls' home."

"That was your choice, Rachel. We didn't ask you to leave."

I put my hands over my face and bit my palms for the rest of the ride.

"Do you want me to come home?" I asked when we pulled up to The Harbor.

"What kind of silly question is that?" Mom said. "Of course I want you to come home. You're a very important part of this family."

I thought about the summer we'd gone canoeing on some lake in Wisconsin. I was about eight years old and Dad had put me in charge of four peanut butter and jelly sandwiches wrapped in tinfoil. I balanced all four of them perfectly on the ledge of the boat but at some point the boat hit a wave and the sandwiches went over. Without much thought I jumped off the boat into the freezing cold lake in my shoes and my dress, swimming frantically toward the four floating sandwiches.

"Remember when I tried to save those sandwiches in Wisconsin?"

"Wisconsin?"

"You know, when I jumped off the boat in my clothes?"

"Oh God, that was funny," Mom said. "That was really funny. What made you think of that?"

"I don't know."

"That was a good attempt you made to save those sandwiches." Mom placed her hand on my shoulder.

"I want you to realize that this sucks. It really, really sucks!"

"You're right," Mom said. "This really sucks."

"Do you *really* think that it's my choice to be here?"

Mom had pulled into the alley. I looked through the back window of The Harbor and into the kitchen. "I mean the truth is, I don't mind it here so much. It's better than being at home, but it's not exactly where I want to be, you know?"

"You'll see," Mom said, "things will be different once you're in college. You'll be older. Dad's not going to be this protective forever."

It wasn't Dad's protection of me that seemed to be the problem, but rather his fear of the world and his perception of me in it.

I kept hoping that Mimi would pick me up and drive me home and tell me to REALLY pack my bags this time, that she and Mrs. Pollock had found the right place for me.

"Do you think it's normal how crazy Dad gets around me, Mom?"

Mom rubbed my shoulder as I sat in the passenger's seat, staring out the window.

"Would you like to come to Cancún with us?" she asked.

I wanted to get through the remainder of high school without going home. I would have happily agreed to live with friends or family or in a permanent type of group home. But I wanted it without the repercussions of asking for it, without humiliating Mom, Dad, or myself.

"Cancún? That's what you're thinking about, Mom?"

"I think I hear you turning angry right now, Rach, and I don't want to be talked to in that angry voice."

I did want to go to Cancún. To close my eyes under the powerful warmth of the sun, to run barefoot down the beach for miles until I hit Club Med, shoot tequila from the bar on the credit card of some kid's parents.

"Who cares about Cancún right now?"

"I do," Mom said. "I love Cancún."

And so would I.

Despite The Harbor, and my dread that nothing in Dad would change, and how painfully long the next two and a half years were beginning to feel, I loved the hell out of Cancún that year. The second night, I'd lie passed out on the floor of the bathroom, my cheek in a pool of vomit, my body exhausted from retching, and I'd think, despite feeling awful, it was a good thing to be getting away with. Good to be taking advantage of the freedom I had when I had it. The next morning Dad would ask if that was me he'd heard retching. I'd say yes, but at least I got back to the condo by curfew. He'd advise me not to mix liquor and beer, and I'd thank him for the advice, and hope he wouldn't use it against me later. Then I'd grab a piece of fruit and sit down by the ocean, surrounded by a bunch of Americans slathered in suntan lotion. We lived in such close proximity to normal, it would be easy enough for me to close my eyes and believe, if only for a couple of weeks, that we were a part of it.

"I think we could all use a little vacation," Mom said. "Let's get you home for Cancún."

■

Each night of Hanukkah I took the little menorah into the bathroom and lit the candles, while the girls sat downstairs watching TV. I didn't want to explain that there were people in this world who didn't celebrate Christmas. Plus, lighting fires in The Harbor was not allowed. On the floor, wrapped in a towel, I'd quietly say the prayers. After the menorah was lit, I'd place it on the sink, where I supervised the burning as I shampooed my hair.

Dad was at his best in synagogue. He seemed to relax. I could see it on his face and in his body, which sank into a comfortable slouch. He almost always shifted into a good mood by sundown each Friday. He was sold on the idea of being a good Jew, of teaching us how our history made us who we were. I was sold on other people's impression of Dad as a good Jew, fascinated by the extremely different persona he displayed in public. He cracked jokes with the security guards who sat by the entrance doors of the synagogue, he shook hands with the men distributing prayer books on our way into the sanctuary, he made a point of complimenting the Rabbi's sermon. There was no doubt we appeared perfect on holidays and Friday nights. We must have, because even to me, we felt right in the eyes of others.

Jenny complained about going to synagogue, and I don't think Mom minded much one way or the other, but I was almost always eager to go, if only to enjoy Dad's good mood, and this bonded me closely, if only for a night, with him.

He called on the fourth night of Hanukkah. I was sitting at the kitchen table with Kimmie, making a Christmas ornament with tinsel and glitter and Styrofoam balls when Fat Macramé waved me to the phone.

"How are you, Dad?" I asked, wishing I'd let him ask me first.

"Pretty good," he said. "Can't complain."

"That's good."

"And how are you, Rach?"

"Fine." I stared at the back of Kimmie's head. "I'm fine."

"Good, good," Dad said, though I couldn't tell what he found good about it all.

"Do you want to come home soon?"

"Sure."

"Are you having a good time over there?"

"It's fine."

I stood with my back against the wall, staring at the food request list that was taped to the refrigerator. Someone had written HOME RUN PIZZA three times in a row.

"Rachel, you still there?"

"Yes."

"We miss you."

I'd never heard him say it quite like that. It was as if he'd been saving up to tell me all his life.

"You do?" I asked.

"Yup."

His voice quivered nervously.

"So we'll see you soon?"

"Okay, Dad."

"Okay," he said. "I love you."

A deep embarrassment filled me. It was not just Dad saying he loved me but the discomfort I detected in his voice upon saying it. I'd grown more comfortable with Dad telling me I was scum than expressing his love, and his sudden affection brought on a shame similar to what I felt when strange men on the street whistled at me. I looked straight out the window, at the purplish shade of a bitter Midwestern cold, hoping to never be on the receiving end of those words again.

The Saturday before I went home we took a supervised trip to the strip mall. Our first stop was the roller-skating rink, which seemed to excite almost everyone but me. I leaned up against the wall and watched the girls skate by in circles. Afterward we were given $10 to spend at Kmart. I was determined to talk to the small Asian girl who slept in my bedroom before I went home. As the girls ran wild through the parking lot, with Fat Macramé walking slowly behind, I sidled up to her.

"What are you going to get?" I asked.

She had huge brown eyes and when she looked at me I realized that she wasn't all that shy in the same way I wasn't all that shy. She just didn't need to talk all the time.

"Underwear, I think."

"For you?" I asked.

"Yeah," she said. "You got a boyfriend?"

"Yeah. Well, not really."

"Because you could get him silk boxers."

"Yeah," I said, "I could."

"Or cologne."

We walked through the sporting goods section. One of the other girls had picked up an orange cone and was using it as a megaphone.

"Do you have a boyfriend?" I asked her.

"No," she said.

The underwear was sorted by size and thrown in piles in big plastic bins. We rustled through the fabrics, tossing the good ones on top. Mom, Jenny, and I shared the same pairs of underwear and bought them in bulk. I had never bought stuff this nice.

"You like this pattern?" I held up a gold-and-pink-paisley pair.

She seemed to approve. I found two pairs, for her and for me.

"I got a baby growing in me," she said, smiling.

I looked down at her tummy. She was shorter than me and smaller, even with the baby. I fished through more underwear, thought about which bathing suit I'd bring to Cancún.

"How many months are you?" I asked.

"Three."

"Three," I repeated.

"I was living with my aunt and uncle," she said.

"Oh," I said, nodding like I understood whatever it was she was trying to tell me.

I watched her face change; her eyes went from unknowing to knowing, and it made me think of Agnes, how she killed that baby and convinced herself she hadn't.

"You want to keep it, definitely?" I asked.

"Oh yeah, I want it and all. You get kicked out of your house?" she asked me as we waited in line to pay.

"No, not really."

"You do drugs?"

"No."

"What did you do?"

"I just get into trouble," I said. "With my dad."

"You're pregnant?"

"No." I laughed. "Not like that."

I felt sick. The suggestion, if that's what it was, made me feel lucky as hell for my type of trouble. I had the privilege of education. Confidence to get myself in and out of anything, a privilege the girls at The Harbor didn't have. A privilege that had me indebted to Dad.

The cashier line was long, and I stood in back gripping two pairs of fake-silk underwear. The rest of the girls were outside smoking cigarettes. In the van we showed off what we'd bought. Underpants were placed on faces and thrown around the van, along with leftover Halloween masks, stolen candy, and magazines. Souvenirs, I thought, packing up the underwear with the rest of my stuff.

■

I don't remember leaving The Harbor. Fifteen years later I called Mom to see if she could fill in the details. I was sitting on my couch in New York City. Mom was reluctant to talk about it. She kept on saying, "That place wasn't for you," as if my going to The Harbor had been like choosing the wrong type of sport to play.

Mom says she picked me up on a Saturday and brought me home. She remembers she and Dad followed Mimi to The Harbor the night that I left. I have no memory of them there. She says Mimi was at our house for a long while discussing all our options. I just remember packing my bag and leaving. Mom says Mimi proposed the option of staying with Aunt Jo Ann and Uncle Arthur, or a family friend, but she and Dad felt it was inappropriate to air our private family business.

"But that place was wrong," she said. "We didn't know it would be like that."

"Like what?" I asked.

"Well, by the time we'd made the drive it was late and we'd already decided it was the best option."

Mom says we talked every night on the phone. I only know of that one conversation I had with Dad. Mom remembers sending money to "those people who worked at that place," which would account for the money I had for the bus and for lunch every day, and for the money I got at Kmart, which I had assumed came from the state. Mom remembers how all those girls were smoking and how they were very different from us. "We didn't expect it to be *like that*," she says. "I drove out there a couple of times. Remember I brought you that outfit for the speech meet . . . "

She interrupted herself. "That place was really horrible."

I walked over to my living-room window, looked as far down Eighth Avenue as I could.

"I don't remember it being that horrible."

I remember, for a long time after I left The Harbor, trying to imagine what became of the girls. I remember feeling like I could be comfortable anywhere. And I remember thinking I needed no one. That I was good at needing no one, like some girls were good at violin, a temporary state I mistook for a skill I could perfect.

■

We'd gone to Athens one summer when I was thirteen. I'd gotten lost on top of the Parthenon. I was scared but calm, and for the most part certain that I would not die, even if I never found Mom and Dad again. I knew I was not unsafe. The American Embassy was close and there were herds of police and tourist buses. I was less frightened of being alone than I was overwhelmed by the looming freedom of not being found. I walked around the parking lot, got on and off the tourist buses, scanning for their faces, hoping to hear English. Then I returned to the place I'd lost them. I cried briefly and hard, realized crying wouldn't help, and began to think practically about places I could sleep that night. The sun was setting by the time we finally found each other.

"How selfish," Dad said. "How unbelievably selfish, to run off with my camera."

"Dad," I said, "I didn't run off. Why would I run off *here* of all places? It's not like I speak Greek!"

Mom reached out for my hand.

"Ellen, don't do that," Dad said. "She cannot be rewarded for abandoning us."

I don't know how we found each other, just that finding them didn't feel much different than losing them had. It was in that same way that I returned from The Harbor.

Reconciliation between parents and children was the system's first choice, especially with a situation like ours. And so I returned home, on the condition that we would get family counseling. Dad agreed to go, and that alone seemed like an incredible feat.

Dad treated me like an old friend coming back from war, and I couldn't help but think that I'd earned his respect for toughing it out at The Harbor. Perhaps it wasn't respect but satisfaction he felt in my return. Maybe he thought I'd finally appreciate him after The Harbor. Dad often asked me how I'd like to stand in a three-hour line at Cook County Hospital waiting to go see a doctor. It was his way of showing us how lucky we were to have health insurance, and I could only imagine that The Harbor was his way of reminding me how good I had it at home.

They asked me nothing about my time away. It was as if it had never happened. That killed me, and for several years after, on the way to the airport or Wisconsin or anywhere else that brought us in the general direction of The Harbor, I would loudly announce that we were close to The Harbor. Never to Dad and always to Mom, because it was Mom who had let me go and Mom I wanted to punish. As far as I knew, Dad needed me to go, as much as I needed to leave.

Soon after I returned from The Harbor, Dad and I went for a drive. It was a school night and the roads were extremely icy. I looked out the window. We ran a few errands, listened to NPR. As we neared the house, I told Dad that when I grew up I wanted to be a lawyer. This was a big deal, since I'd been telling Dad for years that I wanted to be a rabbi, which was a lie that Dad loved to repeat to family and friends. He asked if I knew how many Armenians had been slaughtered by the Turks in 1915. I said I didn't. He told me how many. I said, "Jesus Christ. Exactly the kind of thing that makes me want to be a lawyer."

"You want to be a lawyer," he said, "be a lawyer."

"It's not that I don't want to be a rabbi, but I think I'd make a better lawyer."

"Be a lawyer," he said.

"Do you think I can?"

"You want to be a lawyer, be a lawyer. You can be a lawyer. You can be whatever you want to be."

I kept pushing on, since his mood seemed good and I had an agenda. I needed to know exactly how much longer I had in that house until I got another break.

"I've been thinking about this summer, Dad, about doing something interesting."

"You're thinking about doing something interesting?"

"I've done a lot of research."

"You've done a lot of research?" he joked.

It was as if he was half there in the car and half in another world. He seemed to become almost sedate as we pulled up to our house. Sedate was our favorite way for him to be.

"There's a six-week bike trip I'd like to take."

"Six weeks," he said.

"In Ireland."

I remembered just how good it used to feel sitting in the back of Dad's car when I was little, falling asleep before we got home, getting slung over his shoulder and carried upstairs to bed.

"So she wants to ride her bike across Ireland," he said.

"She does."

"All the ways across the hills of Ireland."

"That's right."

"Well, we'll see about that, Rachel. We will see."

But I could tell from his voice and the gentle lazy way he repeated his words that in no time again, if only for the summer, I'd be free.

Only twenty-two more months. There was nothing about twenty-two months that I could not endure.

TEN

Morrie was everything I imagined a private practice psychiatrist to be, white-haired, refined, and expensive. I know, because I paid. Dad said I had the problem, so I'd pay for the help. I fought it at first. I said it wasn't my problem. Whose problem was it then? Dad wanted to know. Was it Mom's problem or Jenny's problem? Was it his problem? I told him it was our problem, but this didn't fly. "Wrong, Rachel, it's *yours*. You're a sick little girl."

Sickness cost me $75 a session for the four of us to go to family therapy, but Dad had had enough after three visits.

Morrie lived with his wife in South Evanston, close to us, and close to the middle school. His wife was also a psychiatrist. They had two grown-up daughters. Morrie saw clients in the living room, which was full of plants and paintings, and antique lamps. He was always dressed neatly, in well-fitted slacks, and he sat with his legs crossed in a chair across from the four of us on the couch. There we sat talking about our house rules. Dad took pride in his work, and just to insure the binding nature of these rules, he had Mom, Jenny, and me sign at the bottom, like a contract. Then he

RULES

1. Friday night observance of Shabat will be maintained. Exceptions will be made only for those special extra curricular activities in which the children have an important commitment, such as a play rehearsal or a school work meeting. No exceptions will be made for parties or other social gatherings even if they are school related.

2. Weekday curfew will be 6:00 PM for both children unless the children are involved in the above mentioned extra curricular activities that make it necessary to attend a meeting later than 6:00 PM.

3. Curfew on Saturday night (when attending a party) will be midnight for Rachel and 10:30 PM for Jenny.

4. The children may attend no more than one sleepover party per month. On the day of the sleepover, they may leave the house at 6:00 P.M. and must return by 9:00 A.M. (Jenny when Sunday School is scheduled) and 11:00 A.M. (Rachel). on Sunday.

5. The childrens' fingernails will be kept trim (not to extend more than 0.5 cm from the cuticle corner); clothing will be appropriate for the occasion; torn or sloppy clothing will not be worn to school.

6. Phone calls will not be made and the phone will not be answered until all homework is complete and the children have taken the responsibility to ask a parent to proof read any major papers or essay questions that are assigned.

7. Any showers taken in the morning on schooldays must be completed by 6:30 A.M.

8. Weekends: Saturday may be used to go to the library or to work on homework at home. All homework must be completed before going out. If work is to be done at the library, an outline of the exact work to be accomplished must be completed before leaving the house. After completion of homework, up to 6 hours (Jenny) and 8 hours (Rachel) may be used for socialization time (day or evening). Sunday is a day to relax and to accomplish: attending Sunday School, reading the newspaper, practicing the piano, doing household chores, working on homework, reading quality literature, listening to music, thinking, dreaming, and even creating.

9. Absolutely no makeup is allowed at any time except during a play performance.

10. Everyone will wake up in the morning to National Public Radio in order to keep well informed.

11. All family members will "pitch in" with the household chores. Areas that need to be covered by all family members are the following:

a. <u>Dishes.</u> Everyone will wash one's own dishes during meal hours; on even days Rachel will wash the dinner dishes, and on odd days Jenny will wash the dinner dishes.
b. <u>Inga.</u> Inga will be fed, watered and walked by Rachel on odd days and by Jenny on even days.
c. <u>Garbage.</u> Rachel will be responsible for bringing garbage to the street on Sunday night and Wednesday night. Jenny will return garbage cans to their original places on Monday and Thursday after school.
d. <u>Bedrooms.</u> Each individual is responsible for his\her bedroom; all items must be put away properly.
e. <u>Help.</u> When asked to help with an extra task, the child will do it regardless of whether she wants to.

<u>FAILURE TO COMPLY WITH THESE TEAM RULES WILL RESULT IN PUNISHMENT AS FOLLOWS: THE CHILD WILL NOT ATTEND ANY ACTIVITIES DURING THE ENTIRE NEXT WEEKEND REGARDLESS OF THE TYPE OF ACTIVITY.</u>

had the rules photocopied, laminated, and duct-taped to our walls.

I'd spent the entire two weeks in Cancún grounded for coming home twenty minutes late, which actually hadn't turned out so bad. The four of us hung out in the condo reading books, eating chips and guacamole, playing board games. It was the case when I got grounded, which was more often than not, that Dad, who made a point of banning me from television, radio, and social outings, would do something kind and unexpected, like take me for pizza or to the movies.

Morrie told Dad he knew how difficult it was to watch us girls grow up. He told his own stories about raising two daughters who spent their teenage years trying their best to sneak around every set rule. "Steve," Morrie said, in a firm yet sympathetic manner that made all his words sound like fact. "This is normal stuff. Kids are going to do what they can to get away with things."

"Then they pay the price," Dad said.

Morrie took some time explaining that there were certain circumstances, accidents, or unexpected occurrences that had to be taken into consideration, like flat tires, like getting lost or caught in a storm.

"I remember this all too well," Morrie said, leaning in toward Dad. I guessed he was trying to get through with some fatherly bonding. "Things happen sometimes, unpredictable things come up. Exceptions always need to be made."

"Exceptions," I said. "We don't believe in those."

"No," Dad said. "We don't."

It was the first thing we'd agreed on in a while.

But I made a lot of exceptions, against my better judgment. I hitchhiked in order to make curfew, thinking my intuition would let me safely navigate any situation. A ride with a stranger was no doubt less risky than an angry Dad, and I figured if something bad were to happen on that ride, at least I'd be spared the trouble for my lateness, assuming Dad might respect my resourcefulness.

Of the several rides I took with strangers, there was one in particular I remember. I was fifteen, and had been waiting at the bus stop for half an hour when I realized it was six o'clock and I was late. It was dark and cold outside, and at that point a ride was my only option. Dad had banned Mom from driving me places. I phoned her anyway but the answering machine picked up, and explaining lateness didn't excuse lateness.

Across the street from the bus station there was a Walgreen's. And inside the Walgreen's were a host of characters in line at the checkout. I watched a tall black man joke sweetly with the cashier. An older man with a younger face, he paid for the *Chicago Sun-Times* and a bottle of Phillips' Milk of Magnesia, both of which made me think he'd be a safe ride home.

I followed him out of Walgreen's and watched him put his key into the door of a little white car, then asked if he was going north. He laughed uncomfortably, said, "Yes I am." I asked if he would drop me off a couple of miles north. That he was driving to Wilmette made him seem even safer. I buckled up in the passenger's seat.

I thought about the couple of times people had shown up at our door with the running-out-of-gas story asking for money, and wondered if he thought maybe I was the one who was dangerous.

He asked if I was far off Dempster. I told him not so far. He said it was dark and if I was close that he would drive me home. I thanked him and explained it was a bit of an emergency, which it was. If Dad got home before me I'd pay with my week-end, and that I couldn't afford.

"My grandma's in the hospital. We're going for a visit at six and of course . . . *who's* running late?"

I laughed to indicate my lateness was typical. I laughed so he'd realize I felt safe in his car.

"Sorry about your granny," he said.

I regretted lying but needed the man to drive faster, so I could intercept Dad. I tapped my fingers against my leg, feeling very much like I was in a video game I couldn't control. It crossed my mind that we could pass Dad in his car, that maybe it'd be faster if I got out and ran full-speed. I looked at the man, he still looked safe; I thought of the Milk of Magnesia. He turned on the radio and to my relief asked nothing more about Grandma's condition, which was good, because I lied so often in this particular way that I quickly forgot the truth. I spent that entire ride staring out the window, glancing at the digital clock on his dashboard, thinking and really believing that if I didn't get home soon Grandma was going to die.

I had him drop me off a block away from home. I thanked him. I told him he saved me, which must have made him uncomfortable, because he felt the need to give me a mini-lecture I didn't have time for. It wasn't smart of me, he said, even in an emergency. But he understood, right? I did it for Grandma.

"You're a nice girl," he said. "I'm just saying, you know, what if I wasn't me?"

"I know," I said solemnly, because he was right.

I continued to hitch rides, from cafés and comedy clubs, parties and movie theaters. I trusted strangers.

On our second visit to Morrie, we went through the rules

one by one and worked in some exceptions. I brought up prom, even though it was two years away. Prom, Morrie agreed, was a perfect occasion for an exception. As far as Dad had always been concerned, exceptions were simply ways for me to get away from him. Out of the house, over to friends, late at school, part-time jobs, summer trips. As far as I was concerned, he was right.

I continued to test Dad's concept of black and white. I lived in a world of *what if* and imposed it upon us all.

"What if I married a non-Jew?" I repeatedly asked Dad.

I'd just met Shawn. Shawn was in boot camp at the Great Lakes base in North Chicago. He came from West Virginia, one of twelve children. I could barely understand him, with the fringe he put on words. We met at Six Flags Great America. I picked him up in uniform. "You're dating a soldier," Nathalie said. "He's in the navy," I said. "Same thing," she said. Shawn was an easy secret to keep. He didn't have a telephone to call me from his base, and I didn't have a dad who let me answer. He could barely get away from boot camp and I could barely leave my house. Still I liked the idea that someone out there was thinking about me. We talked once every couple of weeks from pay phones and saw each other sporadically on Saturdays. One night he bought me an armful of helium balloons, which I proudly paraded around Chicago. Another time he brought me roses, three long-stemmed from the White Hen Pantry, each encased in plastic. My favorite night with Shawn was at the beach around the bend on Sheridan Road, where we rolled up our pants and waded into the water until our jeans were drenched and hanging heavy. It wasn't warm and we were in sweatshirts. He laid his on the ground and pushed me down and got on top, and we went at it like that in our wet jeans, never removing anything, just grinding our bodies into one another until we were both breathless and exhausted.

I never thought to marry Shawn or do much other than what we did, and eventually he moved back to Virginia.

Mark would come next. He was Puerto Rican, said he was Italian. He was adopted. In seventh grade he had stolen his father's car and driven halfway to Texas, where he was arrested and thrown in juvenile detention. He stayed away after that, for two or three years, and when he returned he seemed handsome and changed and deeper than most boys I knew. I guessed that he had seen things. Mark was tough and held his own. The guy who wanted to join the firefighters but never did; the guy who liked the idea of having a girl but couldn't keep to one. My old man this and my old man that. He was always saying "my old man." I wanted him with the distance and longing that one might want an unaffordable jacket, with the hope of owning something that would make me feel outside of myself.

"Okay Dad, what if the hypothetical *we* were really in love and raised our children Jewish."

"Don't ask *what if* . . . " Dad replied. "You know."

"But tell me," I'd say. "Just tell me again."

Almost always this talk happened at the dinner table, after a history lesson on the mass slaughter of Jews during the Christian crusades. Without fail I'd look at Mom, who'd shake her head at the monotony of this conversation. And I kept on looking at her while I questioned Dad.

"So you wouldn't come to the wedding?"

"A Christian wedding?"

"No, a Jewish wedding or a justice of the peace."

"No, I wouldn't come to the wedding."

"Would you disown me, Dad?"

"Oh c'mon, Rachel," Mom chimed in.

"If you married a Christian?" Dad asked.

"Or a Muslim or a Hindu, you know. Would you?"

"Absolutely."

"Absolutely?"

"Absolutely."

It broke my heart and freed me all at once. It was a very loose understanding I formed around commitment, or anything I saw as binding. In my mind, knowing I could lose him so easily made me think everything could be undone.

I lied and often believed my own lies so I could press harder against Dad's black-and-white worldview. Living, or at least fabricating a circumstance that existed in between boundaries undefined. One night, I went to downtown Chicago to get pizza with friends. There was a long line at Giordano's. I looked at the clock, knew I wouldn't make my curfew, said nothing. I enjoyed my pizza, the night with my friends, and then, from a pay phone outside the restaurant, called Mom and Dad to tell them we'd been witness to an elderly woman's purse getting snatched at Giordano's. We were at the police station. We were giving our testimonials. I would be half an hour late. Dad wasn't having it, and not because he didn't believe the lie—I don't know if he did—but he kept on repeating what we all knew: "Your curfew is twelve. There's no excuse for being late. You're gonna pay the price."

I got in the car. I told my friends it was fine about curfew. My Dad was feeling flexible. They dropped me home. He and Mom were sitting on the couch. Upon seeing them, I actually felt like a hero. Like I had spent the majority of my evening at the police station giving an in-depth description of the man who snatched the purse, an exact play-by-play of what had happened, in my eyes.

"Sorry," I mumbled and with confidence, because I'd been a hero, plopped down on the couch.

"You're grounded," Dad said.

"C'mon, Dad," I said, in a voice that indicated I wasn't going to fight him. I'd spent hours in the police station. In fact I'd traded my own Saturday night to sit in the police station, and now Dad was going to punish me?

"She was old," I said. "And she was upset. You know how old people get about their stuff."

Dad watched me carefully. And I knew he couldn't tell what was true and what was false. And to a certain extent, none of us could.

"That is sad," Mom said.

"It's just sad to be that old and that helpless."

"Where was her purse?" Mom asked.

"Back of her chair."

Lies came so easily. And still they won me no exceptions.

Dad looked at his watch. "Twelve thirty-five."

"Dad, can I ask you something?"

"Sure."

"What would you expect me to do? Leave the scene of the crime?"

I wasn't salvaging my weekend. I was used to being grounded. I looked at Dad, who shook his head. "Don't start, Rachel. A rule's a rule."

I wanted to know Dad was there for me. That if something happened, something that threw off Dad's worldview, he might still be there. But the older I got the less possible this seemed.

While we were on vacation in Cancún, Dad's camera had been stolen. He'd been on a crowded city bus and the kid who stole it jumped off and ran away. Realizing what had happened, Dad ran after him, knocking people over, cutting through restaurants and alleys until he had the kid down to the ground.

Dad loved telling this story. His knees were bruised and bloodied and he'd not only gotten his camera back but he'd gotten the kid arrested. Dad kept saying, "I humiliated the hell out of that kid. Humiliated him." Dad hit him a couple of times and then the police were on the scene. "You know how much they appreciated what I'd done?" Dad said. "Cancún's dependent on tourism. Things like that happen, it ruins their entire economy. They were so thankful. So glad I'd chased him down."

Jenny rolled her eyes.

"I'm serious. You think I'm kidding? That kid's gonna pay the price too. Prison system's brutal down here. Brutal."

Mom brought a bowl of guacamole out and set it on the table.

"You think it's funny, Rachel?"

Dad picked up his camera. "You see, I got this back."

"And?"

Dad began to pace around the room. "You remember when your wallet got stolen?"

My wallet had gotten stolen the year before, when I carelessly tossed it in the small pocket of my backpack. Like Dad, I'd been on a crowded bus, and like Dad, I had felt the man unzipping the pocket of my backpack and then it was too late. He'd moved through the bus and jumped off. I got off at the next stop.

"Dad, you're not blaming Rachel for her wallet," Jenny said.

My body got warm. Jenny was fighting for me.

"Of course I am," Dad said. "Because what does she have to

show for it? Nothing. Nothing. Even as the victim, you're guilty until proven innocent."

I began to walk out of the room. Dad told me not to leave, that there was a lesson in all of this. Didn't I want to learn a lesson? Or did I want to go through life being a victim?

"Tell us more about the chase," I said. "Did you knock down any tables?"

Jenny laughed.

"Not funny," Dad said. "And yeah, I did. Things were flying everywhere. You should have seen it. You would have been proud."

I would have been proud.

"Now," Dad said, sitting down in a chair, crossing his arms. "Take rape, Rachel."

Jenny let out a noise from her throat that sounded like boredom or disgust, I couldn't tell which.

"Let's take rape. Girl's guilty if she can't prove that it happened. Right?"

"Wrong," I said. "Other way around."

"Yeah, Dad," Jenny joined in. "That's in England, not in America."

"The U.S.," Dad corrected. "We live in the U.S. C'mon, Jenny, you know that. Now Rachel, think about this. You get raped, you can't prove it, nothing happens."

Jenny's eyes narrowed. She took a step back and took in Dad. I watched her watch him.

"What does this have to do with anything?" I asked.

"That's the way the world works."

"Your world, Dad!"

Dad smiled. "You don't know what's out there."

"Why are you telling me?"

"Because I want *you* to *understand* the way the world works. Because I want *you* to know that your actions have consequences."

Perhaps it was the only protection he could offer. I knew he didn't want anything bad to happen to me, but more than that Dad didn't want me to get into a situation that would require him to revise his theories.

"You've got to consider these things," he said. "You've got to watch who you smile at. Think about what you wear. Do you know what I'm saying?"

I knew, because he'd said it to me several times before.

"You're not a kid anymore."

This I took as the warning he delivered and the favor that he asked: that I should please not be on the receiving end of anything that would jeopardize or complicate his simple and compartmentalized rules of living.

"Because if you don't have the proof, the world's not gonna believe you."

"What did you just say?" Jenny asked.

"That's right!" Dad said. "Guilty until proven innocent."

"I can't believe you think that!" Jenny yelled.

It was not only that Jenny was standing up to Dad; she was doing it so clearly in my defense. Under her breath, she mumbled "hypocrite." She didn't take her eyes off him.

When the conversation turned to rape, Mom had brought more food to the table and announced that she was going to prepare a little feast. We sat down to eat. Jenny and I didn't speak a word.

"I hope you got the point, Rachel," Dad said as I cleared my plate.

I wondered if Jenny felt like I did. If she felt Dad could not offer his protection or if, because he'd specifically been addressing me, she still felt safe.

Our third visit to Morrie was the last. Dad hated the idea of therapy; he said it was like airing our family laundry all around town.

On occasion I'd walk by Morrie's house, hoping to catch him taking in the mail or watering the grass, partly because I expected to catch Mom coming out of his living room (I suspected she continued seeing him on her own) and partly because I believed he knew there was something unfixable in Dad, that rewriting the family rules was like putting a Band-Aid on a broken leg.

ELEVEN

By springtime Mom had lost the weight she'd gained and started reacting to things as they happened, moving around the house as if it was hers again. Dad was in what we called a permanent bad mood; mad at Mom for something none of us could see. Mom laughed it off, saying, "Dad can be in whatever kind of mood he wants to be."

She took Jenny and me out to eat. She let us answer the phone when it rang. We watched movies together. Dad stayed at work late, arriving home after we were already asleep. Some mornings we stumbled past him on our way to the kitchen; he was wrapped in a sleeping bag on the living-room couch.

I knew about the divorce before the papers came. Mom had been dropping hints for months. Finally, in the middle of the night, she crawled into bed with me and asked how I'd feel about moving to Wisconsin. She meant it this time, she said. This time she'd had enough.

"Just you, me, and Jenny," she kept repeating. "We're going to make a life."

I'd been waiting a long time for this conversation.

"I know I haven't been myself these days," Mom said.

"Not your fault," I mumbled, though it was her fault. For not having left him sooner, for allowing him to put her on drugs, for not being the kind of mother I needed her to be.

"It's normal for you to have mixed emotions about me leaving Dad."

My emotions were not mixed. They had been when I was ten. I remember the wish I'd made before blowing out my birthday candles. I'd wished that Dad would not die, but rather would be sent to China for a business meeting and never return to us.

"It's normal for you to have questions," she said, "to feel like it's your fault."

I could tell that she'd been reading books on how to go about having this conversation. The only question I had was whether or not to put my trust in her.

"I know I've talked about this before, but now I have a plan of action. I've done research."

It was as if she was applying for a job.

"And I know what I need to do, Rachel. And I know it won't be easy."

What could be more difficult than staying, I wondered, but sticking around was clearly a god-given strength of hers. With the exception of the fierce and occasional announcement of a new plan to leave, she put all her efforts into staying. She joined support groups on how to make love work. She read books that schooled her on how to gently break the news of divorce to children. I went along with her carefully prepared speech that night, hoping she'd get off the guilt kick so we could nail down the practicalities. Like when we were leaving and how much money she'd need. Like if she wanted me to get a part-time job. Like how often we'd have to see him.

"Are we staying in Evanston?"

"I can't afford Evanston," Mom said, looking sheepishly at the carpet and shaking her head. "But there's a little town in Wisconsin called Wassa."

"Wassa," I said, trying to subdue my excitement.

"You may think that I have not been doing my homework, but it takes a long time to research communities, Rach."

"What's Wassa like?"

"Believe it or not, Wassa's got a decent-size Jewish community. There's a Reform synagogue and a good public school."

"When would we go?" I asked.

Mom needed to save a little more money. She'd only just started looking into apartment prices and didn't want Jenny to know anything about it until she had all the arrangements taken care of. Mom thought she'd be upset starting at a new school. Jenny had a temper and a hard time making friends. Jenny didn't like change. Jenny might have a learning disability. These were all things Mom felt she had to take into consideration.

"I'm thinking December, maybe midyear."

"I think that sounds good," I said, calculating the year and a half I'd have to enjoy life without him before I'd leave for college. Thinking how in Wassa I'd get myself a boyfriend and a part in the school play and maybe a used car.

"I'm confiding in you, Rachel, because I'll need your help with Jenny when the time comes, but also because I need you to believe in me. I'm doing it this time."

"I know, Mom."

"He's been so crazy. Making me crazy."

Mom and I sat in my bed for a while. Then she leaned over and kissed me good-night.

"This time, Rachel, I'm going through with it."

"All right," I said, getting back under the covers.

"This time you will see."

I'd see. Once we were situated somewhere in the middle of Wisconsin, I thought, far away from Dad. I was ready to move wherever she wanted to take us.

The next month was spent somewhere in between Evanston, Illinois, and what I imagined Wassa, Wisconsin, to be. I staged scenes of our new lives: the three of us furnishing a small flat or insulating an old, broken-down barn, the part-time waitress job I'd get to help Mom with the bills. Free dinners we'd eat in the restaurant where I'd work, where Mom would fall in love with the owner, remarry. Maybe he was a divorcé with gray hair and kids our age. Maybe he was a widower.

Mom made a habit of coming into my bedroom late at night when she thought I was asleep. She'd brush back the strings of hair that fell over my face, and talk about social-work jobs she was qualified for in Wisconsin. She talked and talked, sometimes I opened my eyes, and other times I let her think that I was still asleep, but I looked forward to the visits. There was joy in Mom's voice, a certainty I'd never heard.

■

It was around this time that Jenny and I discovered the tapes. It was a rare Saturday night: Mom and Dad had gone out and left us home alone with $20 for a pizza. It was Jenny who first noticed that the safe was open, who called me downstairs, her voice full of urgency. She was already on her hands and knees removing the contents. Her best friend, Ilana, stood timidly in the doorway.

Dad's safe was heavy and made of steel, with an orange combination lock. He'd left it open accidentally, we assumed, though later I wondered whether Dad, like a shoplifting teenager desperate to get caught, had done it on purpose, wanting us to know that he was locking up our past in a secret place that only he had access to.

Jenny and I didn't speak. We both knew we were going in. I pulled down the blinds, double-locked the doors, and set the security alarm. Jenny made her way through the safe, passing me several big bottles of codeine. "Why do you think he locks up drugs?" she asked.

"Probably to knock out Mom."

Jenny pulled out two sets of souvenir coins, as well as a handful of mini-tapes. I climbed up on the safe and took Dad's tape recorder off the bookshelf. Jenny, Ilana, and I sat cross-legged on the floor of the dining room, removing the cassettes from their cases.

The tapes were a collection of conversations Dad had secretly taped over the years. The first tape we listened to was of Mom, Dad, and me in the living room, my freshman year of high school.

"Because you hate yourself."

"Because I hate myself."

"Because you have no self-respect."

"Because I have no self-respect."

I could hear Dad's breath. I could feel him scripting his next move. And for the first time I could hear the sound of my own defeat, the flatness in my voice. Part of me dreaded hearing more, and the other part couldn't stop listening.

"I have homework, Dad."

"Is that a question? Are you asking me a question?"

"No. I'm telling you I have homework."

"Then say it! Don't ask it. Listen to the way your voice rises when you try to say something. Talk. Don't ask. Tell me you have homework!"

"I have homework!"

"Don't yell, just don't ask, you sound pathetic."

"I have homework."

"Ellen. Ellen? Do you hear her voice? That rising inflection? It makes her sound so insecure."

Jenny was on her stomach, laughing at the predictability of these conversations. To me, Dad's recordings indicated a brutal efficiency, a paranoid obsessiveness that made me wonder how far he was capable of going.

I flipped the tape. The words were fuzzy and it was hard to identify when it was from or exactly what was being recorded. "Listen, Rach," Jenny said, her ear pressed to the tape recorder. "That sounds like Nathalie. That's you and Nat on the phone."

After a while, Jenny grew bored listening to my old phone calls. She wanted to hear herself on tape. But Jenny was merely background noise. The occasional question at the dinner table, a cough or a laugh.

Ilana remained silent, picking away at the carpet. We put in another tape.

There were kitchen sounds, the clanking of silverware. This time Dad was mad about my nails.

"See, Ellen, you're making trouble."

"No, no, Stevie. I'm not making trouble."

"Then why aren't you backing me up? A rule's a rule. She breaks the rule, she gets punished."

"I think, because it's the first time, she should get a warning."

"C'mon, Ellen, we've been through this before. It looks cheap!"

"It does look cheap."

"I'm a cheap gal."

Jenny and Ilana erupted in laughter that was both pleasing and wrenching to hear.

"Jenny, are you going to eat your beans?"

"I'm a cheap cheap gal."

Ilana stopped laughing. It must have been an incredibly uncomfortable position for an outsider, but I was glad that she was there, glad that we had a witness.

"Rachel stays in all weekend long. You hear that, Rachel? No going out."

"At all. Whatsoever," I said.

"You being fresh?"

"No."

"You have stuff to take that off your hands?"

"It's called nail-pol-ishhhhh re-mooo-ver, Dad."

"I don't care what it's called, just get it off your hands."

"I'll need to go to the store if you'd like me to remove it tonight."

"Mommy will go to the store. You're not going anywhere."

"Can I go?"

That was Jenny's favorite part, me asking to go to the store. I was partial to the "cheap gal" comment. All of us enjoyed the fact that whole conversations were lost behind the dog's sporadic barking.

I walked around the dining room. Listening to the tapes had the tiring effect of watching a movie over again. It was a lackadaisical attention I paid, as my mind began to fish around for things we could do with the tapes. We had proof.

"Can we take him to court?" Jenny said. "And sue him."

"For?"

"Mental cruelty."

"It's pretty impossible to sue on the grounds of mental cruelty."

Jenny and I had become avid after-school watchers of *Divorce Court*, and knew enough about mental cruelty to know it was never enough.

"Let's play it for Mom," she said.

"Why?"

"So she knows what he's doing."

"Jenny, she's not an idiot."

"She doesn't know about the tapes," Jenny said.

"Trust me, Jenny. Mom's getting us out of here."

"What do you mean?" Jenny asked.

"Trust me."

We put the tapes back in the safe, leaving it open just slightly as we'd found it. After that we didn't know what to do with ourselves. I felt paralyzed in the way one does after witnessing a stranger having a heart attack. The three of us retired to the den for *Saturday Night Live*.

■

One week later we got robbed. It was a Sunday afternoon and we were returning from piano lessons. The robber broke the lock on the back door and came through the kitchen. The alarm Dad had so proudly installed went off, and the robber fled without taking very much.

By the time we arrived home, the cops were already milling around outside with their flashlights, taking notes. Dad went inside, inspecting each room, saying, "The alarm scared them off. See why we need that alarm?"

Mom went upstairs to make sure nothing was missing from the bedroom. Jenny and I took seats on the couch. Each room seemed untouched, nothing missing, except for the safe.

Jenny and I exchanged looks and then sat down in the living room with Dad and the cop. The cop said things that cops say. Lots of burglaries lately, and something about a neighborhood watch.

He was a red-haired man with big arms, and he spoke mostly to Dad but occasionally shifted his glance toward Mom, as if to pay her the proper respects.

"When was the ADT installed?" the cop asked Dad.

"Couple of months ago."

"Well, it did the job."

"Amazing," said Dad. "Grabbed the first thing they saw and ran out."

"He just said that," Jenny muttered to me.

"My guess is they had some knowledge of where the safe was kept, that they went right into your office," said the cop.

"You think it was someone who knew where the safe was?" Dad asked.

"You keep the blinds up or down during the day?" asked the cop.

"Up, usually," Mom answered.

"Down," Dad said.

"Well, anyone can peer right in," the cop said. "That's usually how these guys pull it off. They stake out the house for a few weeks, figure out when you're not around."

"So you think it was someone that knew the house?"

"Could have just been someone cutting through your yard who caught that safe through the corner of his eye, came back one week later and broke in." The cop picked up his notepad. "Can you list the contents of the safe?"

"My coin collection," Dad said.

"Where were the coins from?"

"I ordered them from a catalogue."

"Were they collector's items?"

"Yes. I had a couple of old American coins and a full set of Israeli coins."

"What else?" the cop asked.

"Codeine."

"Codeine?"

"I'm a medical doctor," Dad said.

I had never heard him address himself as a *medical* doctor.

"How many pills to the bottle?"

"I don't know exactly," Dad said, holding up both hands. "Bottles were about this big."

"Those sell on the street for a lot of money."

Dad nodded.

"What else?"

I looked into the cop's face and, for a minute, I felt like Dad was on trial and the cop knew about those tapes and was granting him one last chance to come clean. But then the cop sat back and said, "Well, if you remember anything else, let me know."

I watched the cop's mustache move with his mouth. I wondered what time he'd get off work tonight and thought how much easier everything would be if Dad had a job where he'd be gone all the time.

Did we have maintenance work done? the cop wanted to know. Construction? Gardeners? House cleaners? He wanted to know what our schedules were, and who was home when. He wanted to know if anyone suspicious had been in the house as of late.

Dad turned to me and asked if any of my friends had come back to the house after school. I considered sarcasm, laying blame on Nathalie or one of the kids I went to Hebrew school with, but it seemed about as smart as joking about bombs at airport security, so I just shook my head.

"Are you sure?" Dad asked. "You want to think about that for a moment?"

I looked directly at the cop and said, "I don't believe any of my friends would rob our house."

"But you can't be sure. You really can't be sure it wasn't one of your friends now, can you?"

"I'm willing to bet it was not *a friend,*" I said, still holding eye contact with the cop.

"I'm not sure it wasn't," Dad said. "It was obviously someone that *knew* this house well."

I couldn't get my mind off those tapes. I was certain Dad knew that we knew. Perhaps he'd left it open to test our integrity. Mom and the cop knew nothing. Coins and codeine were a minor loss. For Jenny and me it was not just the concrete and physical nature of the evidence but that Dad had recorded it himself. In the same way he liked to document everything on film, he'd preserved what he felt was his. It seemed confessional and self-conscious of him and it was that which I'd wanted as proof. Instead we sat locked in a silent, shared knowledge of what had been stolen.

After the cop left, Dad put his glasses on and walked around the living room in circles.

"Did you see the way she was sitting, Ellen?"

"Who?" asked Mom.

"What do you mean, *who*?" Dad didn't even say my name; he just raised his chin in my direction.

"I didn't see," said Mom, sounding annoyed.

"With her legs open," Dad said, "with her legs completely open in front of the policeman."

"Dad, are you kidding me?" I asked.

"Am I kidding you? Are you kidding me? You're fifteen years old. You don't sit like *that*."

I looked at Mom for help, but she didn't look like she was even in the same room as us.

"Like what?" I asked.

"With your legs spread."

"Dad, you have serious problems," I said.

"Cheap," he said, shaking his head.

Dad made an effort to correct us every time we used that word for a product. "Inexpensive," he would say. "The shoes are not cheap. The shoes are inexpensive."

Jenny looked at Dad. I knew she was thinking about those tapes, wishing we had rescued them.

"Steve. I didn't notice anything unusual about the way Rachel was sitting."

"That's because you were sitting behind her. Trust me, Ellen. It was totally, one hundred percent inappropriate."

I had no idea what Dad had seen, but I wanted out of the conversation. I wanted out of hearing shoes were inexpensive, I was cheap.

Whatever it was that Dad saw in me, it was something he struggled with, and I hated him for seeing me like that, and I hated myself for being whatever it was he saw. Because neither Mom nor Jenny saw me the way he did, they only felt the weight of Dad's gaze shifting away from them and on to me, where it would stay fixed until I found a way to leave.

Dad walked out of the room. Mom sat on the couch.

"I can't believe it," Jenny whispered, looking directly at Mom. "I can't believe those tapes are gone! I've got to call Ilana."

Later on, we'd joke about it, imagining the robbers listening to the tapes. Later on, years later, we'd wonder if Dad hadn't planned that robbery himself, then, moments after, mildly embarrassed by our thought, we dismissed the idea, agreeing it was just an odd coincidence.

That night, after Dad had settled into his office, I got Mom alone.

"I really don't like being told I was sitting with my legs spread," I said.

"Let it go, Rachel."

"No, it's sick, even if I was. It's sick that he noticed!"

"Rachel, Dad didn't like the way you were sitting. He told you, now it's over."

"It's not *normal* to say things like that."

"We're not perfect. I don't know any perfect families, and I bet if you think about it, you don't either."

"I'm not asking for perfect," I said. "But this isn't right."

There was no right for Mom, just her blind acceptance of our imperfection as "normal" and "human," which allowed her to exist without questioning what was actually happening.

She got up from the couch. Walked over to the chair, picked up the dog's leash and hung it over the banister.

I could feel her silently asking me to drop it, and I wanted to take her by the shoulders and shake her hard until she snapped. I wanted to hear her stand up for whatever it was she thought was right.

Instead she crossed her arms across her chest and said, "You know what, Rachel? We just got robbed."

■

The divorce papers came a couple of weeks later. They were served on a Tuesday night, when Jenny and I were at Hebrew school. Certified and hand-delivered to Dad. Mom hadn't mentioned a thing. The house was quiet when we got home. The lights in the living room were off.

I didn't see Mom coming at me. I didn't see her until I was halfway through listening to the messages on the answering machine. She was standing with her eyes ablaze. I wish I could remember exactly what we said to each other in those moments before we were down on the ground, but I don't even know if words were exchanged. She pulled me by the neck into the kitchen. Then she took me down. My back slammed onto the kitchen floor and all of her weight landed on me, her hands pressing into my shoulders. I stared at the ceiling, because I thought it'd be best not to look her in the eye, the way we were told not to look at a solar eclipse.

"Ya think you're so smart? Think you're a smart little girl, don't you?"

Mom's eyes darted across my face.

"Ya think ya know everything?" she said in some unidentifiable accent she'd adopted.

I thought to respond in a similar accent, but she already had an answer.

"You don't know *nothing.*"

"Anything," I suggested.

"You correcting me, Rachel? You can't correct me, Rachel. You're just a child, a little fucking bratty child."

Mom took a fistful of my hair, picked my head up and smacked it back down on the floor.

I closed my eyes then opened them wide. "Stop, Mom. You're crazy."

"I'm crazy. I'm crazy, huh?"

"Get off me."

She tightened her hands around my neck. It was hard to breathe. Jenny was standing behind Mom, grabbing her shoulders. "Get off her, Mom. Get the hell off her. I'm calling the police."

Our kitchen was covered in 70s-style wallpaper, big brown patterns and pictures that swirled above me. I felt nauseated. Jenny's hands smacked down on Mom's back.

"God, Mom, you can't even feel me!" she yelled, picking up the phone. "I'm seriously calling the police."

"Call them," Mom yelled. "Bring 'em on."

Part of me wanted to surrender; to close my eyes and let Mom beat the shit out of me. I liked to think that in my surrendering, she'd see how defenseless I really was. But I was too stubborn, determined not to let her leave a mark.

"Ya wanna fight, Rachel? You wanna kill me?" Mom asked.

"I'm about to, Mom," I said, realizing that even though I'd taken five years of Tae Kwon Do, I couldn't figure a way out of the headlock she had me in. "I'm about to really hurt you."

"Hurt me, Rachel. Hurt me. Go ahead."

Jenny was behind her again. This time she was holding a large jug over Mom's head, an empty Carlo Rossi jug I'd painted blue and made into a vase. Jenny began to count to three.

"Get off me, Mom," I said slowly, into her face.

"Your breath smells, Rachel."

I almost laughed but Mom was all wound up. We both were. I couldn't feel a thing except for the adrenaline racing through my blood and her hands tightening around my throat.

"You scared? You scared I'm gonna hurt you? Think I'm going to hurt you, Rach? I'm not gonna hurt you."

Once more she lifted my head and smashed it onto the floor. "I'm not gonna hurt you."

My head seemed to be spinning far away from my body, as if it had detached from my mind and was operating on its own. I was flying far above that kitchen, leaving all of it behind, landing somewhere else, years away.

"You're squishing me, Mom," I said quietly.

Her face changed. She closed her eyes for a while and when she opened them again they were wet and full of small red veins. I could see that she'd just come to realize she was sitting on top of me with her hands around my neck. Both of us were short of breath. Mom unclasped her hands from my neck and put them on her hips. A calm, almost childish awe spread across her face. She continued to sit on me, her hands on her hips, looking around the kitchen as if she were a visitor at a foreign museum.

"What are you doing, Mom?"

"I don't know," she said, lifting her body off of mine, crawling away on her hands and knees. "I don't know what I'm doing. I don't know."

Jenny put the jug back on the piano. Mom pushed herself up from the floor and walked out of the kitchen and into the living room. I waited for the slamming of the front door, the screech of her engine as she reversed out of the driveway, but Mom had nowhere to go. She wanted to be in that house. And I would have done anything to leave.

Dad came down soon after. Up from his nap, dressed in sweatpants, he stood in the doorway clutching the *Journal of American Medicine*. He pushed his glasses toward the bridge of his nose and looked at me on the floor. Then he left the room.

I walked into the living room and sat down on the piano bench.

Dad turned to Jenny and said, "What's going on?"

"Mom attacked Rachel," Jenny said.

Dad looked at Mom like she was mildly crazy and then he returned to his bedroom.

That night we skipped dinner. I went to my room and lay down on the rug, concentrating on the dread that was settling over me. It was the kind of dread that never goes away but lags behind, heavy and physical. I felt it in my chest years before I could identify it as dread, and I knew as it deepened, as it became more like a second heartbeat, that it wasn't going away until I went away.

I took off my shirt and scratched my back against the rug until my skin began to burn, until I knew I'd left the purplish-brown marks I needed to feel myself again. Then I curled up on the rug and fell asleep.

I woke up to Mom cradling me, sobbing, holding my head to her chest.

"I'm so sorry Rach," she said. "I'm so very sorry."

It embarrassed me to be in her arms, but I didn't want her to let go. We seemed broken beyond repair, drowning in a distinctly shared shame, and I wanted nothing more than to stay on that rug and have a good, long cry, but as soon as she realized I was awake, Mom moved away, taking a seat on the edge of my bed.

"Dad got the divorce papers in the mail today," she said.

She *had* gone through with it.

"He just went crazy on me."

I sat beside her, and she started all over again with the crying and apologizing.

"I went crazy on you because Dad, he went crazy on me, and I didn't know what to do."

My head was pounding. This was worth it. Mom had gone through with the plan. We were going to leave.

"I know I've taken my anger at Dad out on you, and I know it's happened before. I want to take you to the hospital to get your head checked out. I want to take you to the ER."

Mom reminded me of the way men in the movies repented to the wives they'd betrayed. I let her go on for some time. I felt bad and wanted her to feel worse, though I knew I'd forgive her. By tomorrow even. I couldn't afford not to.

"I feel okay," I said. "We'll go tomorrow."

"This isn't about you, sweetie. This is about Dad, and me, and I'm sorry that I hurt you. When you came in, and I heard you listening to the messages on the answering machine, you just triggered something in me. And I snapped. I just got really angry, things that had nothing to do with you."

I accepted this with a nod, afraid to say too much. The less I spoke the longer she would stay.

She left, returning later that night. Kneeling next to my bed, rubbing my back, bending down, she kissed my hair. I kept my eyes shut as she put her head on the pillow beside me.

"I don't want to move to Wassa," she said. "I like it here. My whole life is here."

■

It was three in the morning, a few nights after the divorce papers came. Mom came to wake me. By then, both of us were so used to the late-night living-room talks that I simply got out of bed and followed her downstairs.

Mom took a seat on the couch. I sat on the orange wicker chair. Dad walked out of his office, divorce papers in hand. He sat next to Mom, perfecting our triangle. The situation had become somewhat staged.

"Did you know about this?" he asked.

"No," I said.

"You didn't know about these?"

"I mean, I knew, but I didn't know specifically."

"So, you did or you didn't, which one?"

"I knew about the divorce, the idea of it, not about the papers."

"You knew *not*? How do you know not? You know or you don't know."

"Fine, then I didn't."

"Not gonna work, Rachel. Tell her, Ellen."

Mom took a deep breath and looked at me hard. "Rachel, Dad and I are a team."

"Right," I said.

Dad put his hands on his head. "Do you think we're stupid, Rachel? Think we don't see what you're doing?"

"What?"

"What do you mean, *what*?" he asked.

"What . . . am I doing?"

"You think you can manipulate your way between Mommy and me?"

"No," I said.

"'Cause you can't get away with it. You're not going to be able to brainwash your mother anymore."

Mom crossed her legs.

"Conniving. That's what she is. Isn't that right, Ellen?"

Mom opened and closed her mouth like a fish gulping for air.

Dad gave a little laugh and shook his head. "My god, Ellen, you know who she reminds me of?"

"Who?" Mom asked, with the kind of obligatory but exasperated boredom that comes from being on the receiving end of somebody else's bad jokes.

"Saddam Hussein," Dad said, sticking out his jaw a bit. "She's just like Saddam Hussein."

"How do you figure?" I asked, folding my hands in my lap.

"The way she works you into believing her lies, Ellen. The way she controls your every move. She's watching you very closely, Ellen. And she's getting to you. She knows your weaknesses. She'll kick you when you're down. Rachel's like Saddam. And you are like the Kurds, Ellen. You're stuck. You don't know where to go next."

At this point I had a decision to make: to get involved or not. Often, I did get involved, not to seriously fight allegations such as my likeness to Saddam Hussein, but to get out of my head and hear the physical noise of my own voice. But it was riskier to get involved. It meant committing to hours of sitting around the living room, punching at pieces of a conversation that had nothing to do with communication or, as I was beginning to realize, sanity or love.

Mom thought conversations like this were absurd but manageable, a low on the scale of highs and lows of her marriage to Steve Sontag. I did not. I thought that these conversations were a huge waste of energy. There were many things I wanted to take part in. School clubs, plays, youth groups, sports, volunteer work. I had a disproportionate amount of enthusiasm toward anything that would get me out of the house. Any part-time job, from scooping ice cream to raking leaves, sounded glamorous. Because I was willing to do anything, I confused my actual interests with potential escape opportunities. I wanted to do everything and anything.

Perhaps that was why I joined every organization imaginable the second I got to college, including things I had no real interest in, like the gay Jewish group although I was not gay, or the rugby team, which kicked me right off the field. I didn't have the ability to distinguish between what was momentarily interesting to me and what I was really interested in. So, getting involved with everything was attractive. It was the act of *not* getting involved that really sucked the life out of me.

This was true with Dad and our living-room sessions as well. It was defeating to sit it out silently, waiting for him to finish, without interruption or protest. Since Mom took no part in protesting, the job of talking back was left to me. Like manual labor, it was work that resulted in a satisfying physical exhaustion. It was an act of endurance I thought would eventually pay off, as I slowly gained Dad's respect as an individual, possibly a leader, definitely someone who would make it in this world.

The idea that Dad secretly admired me for standing up to him opened the door to more benevolent motivations for his behavior toward me. Perhaps he was neither evil nor ill but stubbornly engaged in playing devil's advocate. Perhaps he was overprotective because he was frightened that the world would crush me, that my gender would hold me back if he didn't toughen me up. Perhaps he thought putting me through hell now would ultimately ensure my determination to succeed in the world.

"Dad, can I ask you a question?"

He looked at me briefly, his expression less that of a man than of a seventh-grade boy in the midst of a heated foursquare game.

"Could Saddam and I be like a brother-sister tag team?"

Dad wasn't looking at me anymore. He was looking directly at Mom. He was giving her that look that one gives when proving a point so huge there is no need for words. The look that said, *do you see what this person is made of*?

"I mean, do you think there are physical similarities between myself and Mister Hussein?"

Mom was biting her lips, the corners folding upward into a smile. She was looking away, which meant she was on the verge of laughter, which meant my own self-amusement had leaked out and was infecting her. All of which did little to disprove Dad's belief that I was manipulating Mom into leaving him.

"Do you see the way she is? Sick. It is sick. Do you see that? She cares about nothing."

"Dad, I care about the Kurds."

It was probably a stupid thing to say, but I thought that if I changed the topic to something of greater worldly importance, common sense might strike Dad to snap out of it and realize how absurd he was being. He cared about the Kurds. He cared about what was going on in Iraq. We discussed the Gulf War regularly. Subjects he was rational about, subjects we agreed upon.

"Steve, this divorce *was* my decision," Mom said. "Something I shouldn't have told Rachel about in the first place."

"That's not what we're talking about, Ellen. Rachel needs to know her deceitful games are not going to work. And you need to be the one to tell her."

I imagined Mom as the player in Capture the Flag who never knew where to run, the lonely kid in gym class who made accidental goals for the wrong team.

"Stevie, I will not turn against you."

"Ellen, that's not the point. The problem is Rachel. And I tell you, having this conversation in front of her just gives her ammunition."

"In that case, can I go to bed?" I asked.

"No, no," Dad said. "We're just beginning. Tell me now. Did you know about the divorce?"

"No," I said. "I told you."

"You're a liar."

"I didn't know about the papers, Dad."

"And a cheat."

I had assumed Mom would tell him before the papers came. That, by the time they came, Mom would have found a place for us to move and enrolled us in school. We would have had a stiff but agreeable family meeting at IHOP to work out the details of visitation and child support. We all would have been in it together.

"You're a liar, lying through your teeth." Dad pointed at me and shook his finger. Then he looked down at his finger like it was a malfunctioning toy.

"Listen, I knew about the divorce but I had no idea that there were papers, that they would come in the mail and all."

"That's right, Stevie, she didn't."

"Quiet, Ellen," Dad said. "I've been watching her plot."

"Are we done?" I asked.

"Steve, are we done?" Mom echoed.

But Dad was just getting the ball rolling.

"Stay here," he said and left the room.

I wished I had the guts to leave the room as well. To walk into the den, uncover the piano, and play. Maybe a good sonata would shut him up.

He returned with a legal pad and a pen.

"Number the page," he said. "We're going to play a little game since you seem to enjoy playing games."

He removed a toothpick from the pack in his pocket and carefully picked between his two front teeth.

"Ready to write?"

I placed numbers in the margins like I would have for a spelling test.

"Selfish."

I wrote it down.

"Rotten, worthless, brat."

He was talking slowly, making a point to choose his words carefully.

"That's what you are."

"All right," I said.

I thought about how to arrange those words. There was the option of slotting each adjective near a number and taking up four whole lines, but I decided to conserve space, and write all four on one.

"Say it."

I repeated after him.

"I'm," he said, "say it in a full sentence."

"I'm a selfish, rotten, worthless brat."

Everything inside my chest began to churn. I rested my eyes on a water stain, a ring that marked the wooden coffee table.

"Ellen," Dad said, "are you just going to sit there?"

Mom removed her hands from her knees and folded them across her chest.

"Scum," Dad continued. "You, Rachel, are the scum of the earth."

"It's a wee bit harsh," I said in my best British accent, indicating a wee measurement with my fingers.

"I'll kick you out of here in a second, I really will."

"Please do," I said.

"You can leave when we're done."

He removed his glasses and leaned in. "How does it feel to wake up every morning knowing that you are the scum of the earth?"

A rhetorical question, I decided.

"You're like a snake. You are scum."

"Scum," I muttered.

"Don't just sit there, write it down."

I jotted *scum* on line two, and quickly drew a snake with its tongue coming out of its mouth.

One of my legs had fallen asleep. I lifted my foot an inch or so from the ground to alleviate the tingling dead weight. I lifted it higher and began to shake off the numbness, but the tickling sensation shot up through my hip. I pinched the skin of my leg, which must have grabbed Dad's attention, because he asked if I was listening to him or if I was busy playing with my pants.

"My leg fell asleep," I said, which for some odd reason caused Mom to react.

"Your leg fell asleep?" she asked, looking at the leg, as if this sort of numbness was something she could see.

"Ellen," Dad said. "Why do you respond to things like that? Mommy's ashamed of you. Aren't you, Ellen?"

Mom seemed to be floating off in the distance, like a helium balloon reaching a certain height before being forgotten.

"Say it, Ellen. You are ashamed that she's your daughter. She can't just hear it from me. I don't want to look like the bad guy, Ellen."

"I am ashamed," she said.

But as she said it, she closed her right eye and winked, as if to show me that she didn't mean it, a reminder that this was just a game, a joke of sorts. Together, we were taking one for the team. Mom never seemed to get how real this was, that this was our life. This was our Saturday night sitting around in the living room. Mom thought that, like every other family, we had our problems, that she was caught somewhere in between her stubborn husband and her stubborn daughter.

"You are a dirty worm," Dad continued. "As low as they go."

I thought about the poolside limbo contest in Cancún, how Mom got up to play, how the song "Tequila" was blaring from the speakers, how drunk the kids were and how sober Mom was and how Dad kept looking at her like she was a fool and the announcer kept saying, "As low as they go, folks. That's as low as they go."

"You are a traitor."

"I, Claudius," I said.

Dad didn't like that, but decided to move along, sputtering off a trail of words he seemed to have found in a thesaurus.

Like a coughing fit, Dad started and he couldn't stop.

"Deplorable, contemptible, degenerate, perverted."

"Hold on," I said, waving the pen in the air. The writing was a welcome distraction. "Too fast, you're going way too fast."

He stopped talking and waited for me to catch up.

"Shameful, shameless, revolting, detestable," he continued.

"Dad," I said, placing the pen behind my right ear. "A lot of these are synonyms."

Sometimes I misread him. Sometimes I thought he was going to kill me for being sarcastic, but he seemed to be relieved. Relieved that I was hearing what he had so diligently prepared; relieved that we were getting through the evening. Relieved that even if my attempt to speed up the process and extricate myself from the room was insincere, I was verbally willing to accept what he'd defined as right. Dad sat back against the couch and let me write. Slowly, he repeated all the words that I had missed, and made me read them back to him.

"I don't want a daughter like you," he said.

"No, sir."

"I have no respect for you."

I considered telling him we'd already covered this on various other occasions, but I could tell by all the little creases in his forehead that I wasn't going anywhere until he got it out of his system.

"Let me ask you a question, Rachel."

"Yes, sir," I said.

"Stop it. Would you stop it? Ellen, see how she is? She thinks this is funny."

"Rachel," Mom said, "you're making it worse."

I looked into her eyes, begging her for another wink, something to indicate she knew this was wrong.

I thought of Nathalie's father and how he kissed her head when she came in from going out at night. How he kissed her head because he loved her, but also to check for cigarette smoke, how she knew what he was doing, and how she let him. Dad and I would never have anything that gentle together, but at least he was tapering. Soon I'd go upstairs, close my eyes.

But Dad had one last thing. "I want you to hear this, to really understand this."

He shook his head like it hurt him to say it, like he didn't want to say it, but I had forced him to come to this place and now he couldn't go back.

"I wish you were never born."

It came out almost shyly. And I thought he'd come to a stop, realized he was killing a certain part of me, and I thought that Mom was going to blow the whistle, declare that we'd gone too far. But Dad looked up from the carpet, into my eyes, and said, "I mean that, Rachel. I really do. I wish you were never born. I really, really do."

Everything he'd said up until that point felt like a necessary extraction. Like the relief that came from vomiting, shitting, a simple emptying that Dad needed. Eventually I learned, just by the movements he made with his hands and the tiredness in his voice, when he was coming to the end. Almost always determined by the satisfied look that would come over his face, the same look hungry men get at the end of a meal they can't finish, and with that satisfied look came the knowledge that when he was done speaking we would all find relief.

I couldn't look up, didn't want to cry. Instead I fidgeted with the pen. The ink was running light on the page.

"What's wrong?" Dad asked. "Is the pen not working?"

He stood up from the couch, walked over to where I was sitting. I handed him the pen. He held it up to the light and examined the ink. A quiet came over us. We all just sat there for a while, as if absorbing the last of sun before the brisk air of evening. Dad gave the pen a good shake. He seemed genuinely concerned with the quality of the pen, in the same way he always seemed concerned with whether I truly understood how I got the answer to a math equation. And that's what confused me most, because as far as I could see, that was love.

That night I stopped believing in Wassa, Wisconsin, the same way I'd stopped believing in the tooth fairy. I lay in bed, too angry to sleep, waiting for morning, listening for the sound of Dad's car pulling away. And when it did, I waited for Mom to come into my bedroom, to apologize for another night. I heard her move up and down the stairs. I heard her calling to the dog. Then I heard the door slam shut. I stayed in bed through my first class, and my second. Then I took a long, slow walk to school, breathing in the warm, moist scent of spring, knowing very well that it was up to me to get myself out of the house.

That was the morning I stopped believing in Mom.

TWELVE

Dad was proud of me. He told all his colleagues that I'd been accepted as a congressional page for a semester in Washington, D.C. I didn't think that was his style, but one Saturday afternoon Mom, Jenny, and I went to a barbecue at the house of a woman Dad worked with. Jenny and I were helping ourselves to the buffet when the woman asked, "Which one of you is Rachel? I hear you're off to work for the government. What a wonderful opportunity."

I nodded politely and said, "I know."

"Well, your father talks about you all the time," she said.

"Really?"

"All the time," she said. "Do you want to be a politician?"

"Not really."

"Well, what made you think to be a page?"

In the car on our way home I told Mom I was surprised that Dad talked about me to his colleagues in that way. She said, "What do you mean *in that way*?"

"He hasn't told me once that he was proud."

"Well of course he is. We both are."

I was washed by a slight tinge of guilt for having thought

so little about the Page Program and so much about leaving home.

Mom and I went to a uniform store in Chicago's warehouse district, where I got the required navy blue suit jacket, a gray polyester skirt that fell below my knee, and three starched, white, button-down shirts. "Can you believe this is what you'll wear each day?" Mom asked from the other side of the makeshift dressing room. I looked at myself in the mirror. The jacket was made of heavy polyester, rough like cardboard, with two bright gold buttons on the coat and three smaller ones on each sleeve.

Mom stuck her hand in and wiggled it around. I passed her two pairs of gray slacks. She stuck her face in. "Did you try on the skirt?"

"Get out, Mom." I stood in the suit jacket and my underwear. "I'm naked."

"You're not naked; you're in the suit."

She laughed as I pulled the curtain closed. She opened it again, stuck her face in.

"Look how nice that jacket fits."

"Just let me get dressed," I said.

I saw her dark silhouette on the other side of the curtain, heard her laughing, as I slipped my shorts back on and pushed the curtain to the side. "Mom, I'm not four."

At the register, as the old Polish man wrapped the polyester suit jacket in plastic, Mom took my face in her hands. "I'm so proud of you," she said. Her eyes were wet, and I thought it was from laughing but I wasn't sure. Mom hung the jacket on a hook in the back of the car, with the crisp white shirts. "Are you crying?" I asked as she backed out of the lot. "No," she said. "I just don't know what I'm gonna do without you."

Dad and I got along well that summer. There were little

things, of course. Things that caused him, now and again, to say, "That's it, Rachel. You're not going to Washington." But both of us knew I was on my way out.

A month before I left, my friend Marisa came to visit from Atlanta. I'd met Marisa years before; she was one of two girls I kept up with from the wilderness camp Dad had sent me to.

The second night of her five-day trip, I fought with Dad over curfew. Dad handed Jenny a video to return and money for ice cream and sent her out of the house with Marisa. "Sit down," he said, once the two of them were on their way. "Get yourself a pen and paper."

Maybe it was because I was leaving soon and had less to lose, but I let myself break down in front of Dad. I cried and cried until I heard Jenny's key in the door. Mom stood up from the couch. "They're back, Steve," she said. Dad told me to pull it together. Then he sent Marisa upstairs to wait in my bedroom while he proceeded with the list. When he was through I got up to leave.

"She goes tomorrow," Dad said. "Ellen, you'll call the mother."

"Her name's Marisa," I said.

"And the mother will change her flight and you will drop her at the airport."

In my bedroom, I put my hands over my face and held my breath.

Marisa closed the door. She took me in her arms.

"You don't understand," I said, my body boiling ina sudden heat.

"I heard," she said.

"It's worse than what you heard." I buried my face in her shoulder and let myself go.

"I heard enough," she said. "He's crazy."

I felt my tears and snot and saliva soaking the cotton of her T-shirt. Then I was facedown on the bed, letting her stroke my back. Marisa held my hand that night as I told her everything.

She asked if I would call her mom. I said it didn't matter; I was one month away from leaving. Then she demanded that I call her mom. "Just to talk, just to have someone on your side."

"School counselor's on my side," I said.

"But my parents can help you out. They will, I promise. We did it for another friend, she stayed with us a year, and it wasn't as bad as *this*."

Mom and I drove Marisa to the airport the following afternoon. We both sat in the back. When we got to the airport she handed me a piece of paper with her mother's private line.

"Call her collect. Call her anytime. You'll see, she's good to talk to."

I fantasized about moving to Atlanta and living with Marisa, after D.C. and before college. And though I knew there was no way in hell Dad would consider it, I hoped Mom might ease into my absence and find the days more pleasant without the distraction of having me home. I figured if she wouldn't leave, she could at least help me stay away.

At fifteen, I wasn't thinking about where I'd end up, just the many different ways in which I could continue to escape.

■

We made a road trip of it. All four of us drove to D.C. in the Plymouth Voyager. I don't remember Jenny being with us. She doesn't either. But she's in all the photographs.

What I most remember was Mom, Dad, and me walking around D.C. the day before they left, getting oriented with all the famous buildings and monuments. Dad was in an unforgettable mood, laughing at everything I said. I was walking ahead of them, thinking they'd be in Evanston by the time I woke up the next day, when I heard Mom yelling. "No no, Stevie," she said, "don't you walk away from me."

She no longer saw anything other than Dad, and he was walking away from her, catching up to me.

"You getting excited?" he asked.

"So excited," I said.

"You think you're ready for this?"

"I know I am."

"It'll be an interesting year. A really interesting year," he said, massaging my neck with his hand.

"I'll have lots to report."

Mom was trailing behind us yelling his name. Not once but over and over again, her voice becoming louder.

We pretended not to hear, walking quickly ahead. She ran to catch up, still yelling *Steve* at the top of her lungs. I gave her a look, but she didn't stop yelling. "I don't like being ignored, Steve. I don't like being ignored." To my relief Dad picked up the pace; neither one of us wanted to play a part in the scene that she was making. Dad began to grill me on important dates in American history.

Then from behind us came a high-pitched scream. I turned around. Mom had stopped walking, and, much like a gym teacher about to blow the whistle on a foul, she had her hands on her hips with her knees slightly bent and her feet pigeoned out. She screamed at the top of her lungs, "I give up, Steve. I give up!" Then she ran into the middle of the street with all the cars coming at her. Horns blared. Mom stood there, looking back and forth between Dad and the oncoming traffic. Then she moved off to the side and fell to her knees. Dad's face was twitching.

She'd done this before. Years ago on our way to an opera in downtown Chicago, she opened the door and jumped out in the middle of the highway and ran. We drove around Chicago looking for her, missing part of the opera, and I remember feeling like Dad should have left her there, and hating myself for feeling like that.

My real fear was in becoming like Mom, a dread that began as soon as I saw she'd lost all power in her uncurious love of Dad. Mom's disinterest in questioning uncomfortable territory was the single most frightening quality I could see in a woman.

In making sure I did not become Mom, I became more like Dad. It seemed wiser to take Dad's attributes, and shed them later on, than to be depleted of strength.

Mom picked herself up from the street and made her way over to us. Her face was streaked with tears. For some time we just stood there watching the lights change from green to yellow to red. Finally, we walked back to the motel. Mom's hands were dirty from the street and she tried to take mine in hers, but I pulled away.

"I'm sorry, Rachel. I'm sorry that I did that. I'm just sad to see you go."

I looked at the unfamiliar city behind her; a reminder that I was no longer stuck. She was stuck.

"Okay," I said. "It's okay, Mom."

But I didn't know if she would miss me or if it was the thought of me no longer being a buffer between her and Dad.

I wondered if Dad's anger would soften without me around or if it would simply be shifted onto Mom. I knew it wasn't Dad's anger that Mom feared; it was the possibility that even in my absence she'd continue to be ignored.

Because Dad did not see Jenny, I chose not to see her. And because Dad saw me in the disproportionate way that he did, so did Jenny and Mom. I had defined my own strength in surviving him.

"She just hates to see us go," Dad said jokingly to the security guard in the lobby, later that next day.

"There's always a few of them," said the security guard. "One little girl just broke down, she was so sad to see her parents leave."

"Not this one," Dad said, rubbing the back of my neck. "This one wishes we left yesterday!"

We had a good laugh. The security guard was an older man and I imagined he'd been working there a good thirty years or so. "She's gonna miss you. I can see it."

Dad looked at me then, and I could see both pride and jealousy in him. This was an experience he would have liked to have.

I walked them to the car. It was a bright, sunny day. "We're gonna miss you," Dad said, his hand resting on my neck.

I didn't know what part of him could miss me, but I believed him when he said it.

I'll never forget that first night. I was thrilled to be away from home, and a little terrified that without the challenge of my father I'd amount to nothing. It was not him but the weight he held in our world that left me feeling slightly devoid of purpose once I got to D.C. There was nothing real like Dad had been real, there was nothing anchoring me down. I was eager for my roommates to leave before I'd even met them. After we'd introduced ourselves and claimed our beds and hung our suits in the closet, after they left to get something to eat, and I was finally alone in that room, I locked myself in the bathroom and cried and cried.

THIRTEEN

The rules were easy to follow. We couldn't smoke cigarettes. We couldn't hold hands or go to political rallies or protests. We couldn't drink. We couldn't do drugs. We couldn't wear skirts above our knees, because of previous sex scandals involving congressmen and pages.

The program was made up of fifty-two sixteen-year-olds from all parts of the country, with a loose interest in politics. We woke before sunrise in order to make it to classes, which were held in the attic of the Library of Congress, beginning at six a.m. and ending three hours later, when we reported to the House Floor. When Congress went into session earlier, school was canceled. If Congress stayed in session until two a.m., so did the pages. It was understood that we were not in Washington to be schooled in a classroom.

During the day, I saw many pages briskly darting from one location to another, running errands with the obedience and focus of future senators. There were others, like myself, who took the time to get lost: wandering through the city, stopping to talk with police officers, political protesters, school groups, Soviet tourists, homeless men, then slowly making our way back to familiar territory.

We were sixteen years old and smack in the center of our government, receiving priority treatment we'd lose access to after that year. We sat in the balcony while George H. W. Bush delivered his State of the Union address. We shook hands with Mikhail Gorbachev; we skipped lines at the security gates. We were privileged and full of curiosity, and, regardless of our political and personal differences, we all took pride in the responsibilities we were given.

Weeks after we arrived in D.C., the General Accounting Office released an audit to the public, stating that House members had bounced over $25,000 in checks without penalty. The media demanded the names of congressional offenders. The Speaker of the House defended those involved but with the pressure of the Republican Party, the media, and the public, he consented to full disclosure. Further investigations were made into congressional spending, unnecessary indulgences, waived parking tickets. When I tried to pay for a haircut with a check cut from the House Bank, the receptionist gave a hearty laugh, passed it to her boss, and said, "We don't take these here." Then she made me promise that it wouldn't bounce.

Soon after, the House sergeant-at-arms, who ran the House Bank, was held up at gunpoint, robbed, and shot. The pages formed opinions and made predictions. There was a beautiful solidarity that spread through the program, just knowing how much inside information we were being trusted with.

In the fall the Capitol was besieged with protesters from ACT UP. They drove into D.C. like battalions, made up mostly of men with short-cropped hair. Some came in costume, dressed as skeletons in black T-shirts with white crossbones ironed on, white face paint with black eye makeup. They came on foot and bike and float. They surrounded the Capitol. The idea was to hold the government accountable for neglecting AIDS research. The protesters shackled themselves to the gates of the White House while President Bush vacationed in Disneyland. Fountains were tainted red with dye. Tourists watched in horror. I accepted as many outside assignments as I could, to take in the protests. Mrs. Donnelly, the head of the Page Program, advised us to eat our lunch in the cafeteria. She didn't want us getting mixed up with the performance art of political protest, at least not in our page uniforms. I grabbed a sandwich and sat down on the Capitol steps, hoping for some conversation.

It didn't take long. Young government employees are appealing to defiant political dissenters, the way Jews are to Evangelical Christians. A man sat down, examined my uniform.

"Work for the government?" he asked.

I nodded.

"You know why we're out here?" he asked.

"Health care," I said, "money for AIDS."

"It's more than that," said the man.

I watched tourists taking pictures of the protesters, wondered what countries they came from and what those countries thought of our free speech, such a blunt and physical form of opposition.

"You don't have to work for the government," he said. The man's eyes were fixed on my government ID. "You choose to."

There was righteousness in his voice.

"I think it's important to understand the system you live under before you decide to fight it," I said.

"The system is the enemy," he said. "You're either with them or against them."

The extremity of his statement got to me. I agreed with the

demands of ACT UP but there were more effective ways of making demands, and despite all the noise they were making, Congress considered them more of a nuisance than an actual voice.

"I think we're fighting the same fight," I said.

"I don't see you fighting," he said. "I see you hiding in a uniform, a bunch of little robots doing the Big Man's dirty work."

"Do you think you're being heard?" I asked. "Don't you think the message gets lost in the ridiculousness of the display?"

"They hear us," he said, "more than you realize."

The next day several members of ACT UP dropped stink bombs in the Visitors' Gallery of the House. We shuffled around, ignoring the smell that took days to dissipate.

I called Dad that night, something I only did when I had a good story for him. First I told him of the protesters and then of the red dye in the fountain.

"What a huge waste of time and money," he said.

I thought he'd appreciate the story, having opposed the Vietnam War, marched on Washington, and frequented the peace rallies around Chicago.

I tried to save it by pasting on a moral.

"I guess the less you're heard the louder you have to yell."

"Are you talking about the protesters, Rachel?"

"Yes."

"You think that's an effective way to make a statement?"

"Not what they did. I think that's stupid."

"Then what's so interesting?"

"That they were so desperate to be heard."

"Oh c'mon, Rachus," Dad said, to my relief calling me by my nickname, "aren't there more important things going on over there?"

There were. I'd discovered acid. The drug wasn't any fun,

but it had been an experience, one that made me sympathetic to people who were mentally unhinged but who appeared normal to the world. That's how tripping felt, like a superpower of sorts, as if I was pulling a fast one on the world, getting away with something, becoming better for it. I loved to replay the acid trips in my head while delivering American flags to Congressional offices, wearing my suit and tie. I wondered if that was how Dad felt. Did he realize what a huge divide there was between his public and private persona? Did that awareness make his skull contract? Perhaps Dad's perceptions were as skewed and distorted as mine had been on acid. Maybe he was waiting for his return to the familiar, the mundane. Perhaps Dad was mentally ill. Suffering more than any of us knew.

When I was in eighth grade, he'd rented *Easy Rider.* I don't remember the details of the movie, just that Dad thought it was one of a kind and I knew we were quite lucky to have a Dad that liked this type of movie. "Dad," I'd said casually, as the credits were rolling, "you ever done acid?"

Dad had crossed his arms and looked up at the ceiling, repeating my question in a slow, dazed fashion. "Have I ever done acid?"

He never did answer the question, which was just the type of thing Dad liked to do. Instead he continued to repeat the question, his tone changing slightly every time. "Have *I* ever done acid? Have I *ever* done acid?"

I thought it was his way of saying, "I've seen things that you can never touch, Rachel."

But I wanted him to see that I was not unlike him. I was not afraid to venture into the darker places that most people chose not to see.

I was not so inclined toward drinking, parties, or boys, but there was something else about acid that made me choose it over and over again. Acid made me feel fragile and minuscule. Acid let me feel like a visitor in a world that could barely even see me; my only responsibility was to preserve my sanity, to control my thoughts and emotions. On it, I felt like I could survive anything, and at the end of a trip, every version of reality felt manageable. It was as if I'd returned from hell and the hell on acid made the hell with Dad seem a whole lot easier.

FOURTEEN

Over my winter break Dad took us on a Caribbean cruise. On the last night of the cruise, I forgot my room key and slept on a sofa in the lounge. Dad was furious that I'd stayed out all night, which made for a memorable three-day drive from Miami to Chicago. As part of my punishment, I was not allowed to sleep. Dad kept a close eye on me and every time I drifted he'd pull over the car and repeat my name louder and louder until I woke up. But the real punishment came after I returned to D.C., when I was called out of class and into the school counselor's office.

Dad had sent the principal of the Page Program an envelope stuffed with apology letters I'd written him over the years. The principal had turned the letters over to Pat Caulfield, the school counselor.

Caulfield was a woman in her fifties. She wore small, round glasses and carried a shaggy stuffed animal under her arm when she made the daily announcements, which made her the butt of many jokes. But she was soft-spoken and kind-hearted, and she kept shaking my letters in her hand and saying, "Really Rachel, I've never seen anything like this."

The letters all began with "Dear Dad" and ended in "Sincerely, Rachel." But they were far from sincere. Strategic maneuvers that took the form of long, drawn-out apologies for sarcasm, swearing, eye-rolling. Words I'd traded for a Saturday night sleepover, or a summer away from home. Words Dad dictated as I typed. Words made meaningless; grand, sweeping statements that I knew would appease him. *I'm sorry I muttered such utter filth. Inside, I hate myself deeply. This is why I say these things.*

Dad was particular about my writing in the same way he was about my acting. He wanted me to feel what I wrote. He wanted me to feel what he felt.

"You don't mean it," he'd say upon reading a draft. "I'm not accepting this letter until it sounds real."

And I'd return to the computer, often with Dad's thesaurus, and find a more sincere-sounding way to express my regret.

Mrs. Caulfield returned the stack of letters to the manila envelope and began moving around the room with haste I didn't know she possessed, pulling college admission books off her shelf, talking about the GED. She called advisors with questions about early high school graduates. "Maybe that's what we'll do," she said. "Get you off to college a year early."

She asked me what kind of college I saw myself at: big, small, city, country? I told her someplace far away from Chicago. She photocopied information about Berkeley, UCLA, and the University of Texas.

Later that week she called me back into her office. Pity had bought me another five months in D.C. Phone calls had been made, papers signed, someone had pulled strings, made an exception for the girl from Illinois with the family situation. My extension came through in late January. I was to stay on for the second semester.

Mom called later that night. She had new divorce plans: a good, solid lawyer, apartment prospects in Skokie, money from her cousins. She said that Dad was sleeping on the couch. She let out a long sigh and said, "It's all over, Rachel. We're not going to live like this." I told her about the extension.

The following weekend she and Jenny took Amtrak to D.C. for the first semester commencement ceremony. Dad, still mad over the cruise, had forbidden Mom to come when he heard I was staying another semester. In the lobby of the Best Western I asked Mom about the divorce.

"Thank you for asking," she said. "Dad and I have decided to work things out."

I'd been standoffish all weekend. Eager to show Mom how good I had it without them.

"What do you want from me?" Mom asked. "You get to stay here in D.C."

I wanted *her*. I wanted Mom to be someone she wasn't, to take on a strength she never possessed, to do what I hoped I would have done in her situation. I thought that mothers were naturally inclined to protect their children, and she was failing. I thought she should have expected more from love, and I held her to my own standards of love, which were conceptual and formed merely in opposition to hers and had yet to be tested in the world.

"I never had a father," Mom said. "I want my kids to have a father."

But I think what Mom wanted was a father for herself.

They stayed for the commencement ceremony and left the next day. We waited for the train at Union Station. I extended my hand and thanked Mom for coming.

"I'd like a hug," she said. I did not give her a hug. I stood with my hands in my pockets while she wrapped her arms around me.

It was true they'd made the trip and I knew she'd pay the price for coming. But I would have rather she turned away from me completely than only half toward me.

"I'm doing the best I can," she said.

"Don't tell me you're leaving him, Mom. You don't know what it does to me."

"Like I said, Rachel. I am doing the best I can."

But I refused to believe she couldn't do better. Mom squeezed me harder. I dug deeper into my pockets.

"We came all the way out here. I would like a hug."

I took my hands out of my pockets and placed them around Mom.

Dad had always made a point of distinguishing between wants and needs, and in the hope of moving on I decided Mom was simply a want that I could do without, a luxury item of sorts.

"Thank you," she said, pulling my head into her shoulder then removing herself from me. "That was very nice."

I saluted Jenny, who rolled her eyes and turned away.

Mrs. Caulfield and I met throughout the rest of the semester, researching ways to keep me out of the house my senior year, but only two colleges accepted early high school graduates and neither was a school I was interested in. With only one year left I decided not to do anything to piss off Dad and jeopardize his paying for college.

Plus, Grandma Dot was dying. She had moved in with Uncle Arthur and Aunt Jo Ann. When they needed a break, we moved her bed and toilet into our living room. Dad asked me to come home for the summer and help take care of her. Jenny had been helping with Grandma all year while I was in D.C. and the first thing Jenny said to me when I got home was, "You have no idea how bad it's been."

"I know. She's really sick," I said.

"I'm talking about Dad. *He's* really sick. Trust me, it'll be the worst summer of your life."

It wasn't.

Because I knew what I could live without, I asked for nothing. I barely left the house. Instead I rented movies and read everything I could get my hands on and followed the strict caretaking schedule Dad had made for Jenny and me. Like everything else, Dad saw Grandma's dying as an opportunity to educate us. It was not enough for us to watch her die, to bathe her and change her and take her to the bathroom and feed her food and oxygen and love. Dad wanted us to know how it felt to work a full-time job. When Jenny and I were "on duty," we were "on duty." Jenny proposed that we work in shifts so each of us could have a break, but Dad wanted both of us with Grandma at all times. Jenny tried again the next week, telling Dad how much healthier it would be if she could have a night with her friends once in a while, how it would make her better, more patient, with Grandma. I didn't say a word. Instead I watched Jenny needing and wanting and asking for what she'd never get.

Oddly, I felt okay about being home. I was needed. Grandma was dying and it was the last chunk of time I'd have with her. But Jenny was just fourteen and she'd had a bitter taste of life at home without me. In a year I'd be gone for good. She had years ahead of her.

Senior year, I made myself invisible. I dropped small enough doses of acid to keep me at a distance from everything around me. Acid distorted and enhanced and confused my concept of reality to such a degree that I could spend full days sitting at the kitchen table saying nothing. And that was where I sat. At the kitchen table, enjoying a bowl of oatmeal with Dad, reading the Sunday paper, quietly filling out college applications for women's schools.

Dad had caught a *20/20* on the benefits of single-sex education. There was no conversation. There was no dispute. Dad would pay tuition if I went to a women's college. When I got a scholarship at Simmons, a women's college in Boston, Dad ruled it out saying I wouldn't be able to control myself in a city.

"Just think how far you're gonna get without the distraction," Dad said. We both knew he was referring to men, but he seemed also genuinely dedicated to my getting ahead in the world. He was full of pride when I was accepted at Smith, and I rode on the wave of his excitement through those last hazy months at home.

FIFTEEN

Smith College was a haven. A shelter from reality. A beautiful, ivy-laced campus full of rolling green hills and lakes and old Victorian houses. Candlelit dinners on Thursdays, afternoon teas on Fridays, porcelain instead of plastic, linens instead of paper, women whose hands shot up in the classroom, a prelude to their articulate answers to professors' questions, women who seemed to be born with a certain undamageable pride and confidence.

Dad took out loans. He refused to apply for financial aid, said there were people who needed money and we were not those people. He bought my books and deposited monthly stipends into my bank so I wouldn't have to work. I was lucky to be there. I knew that, and I wanted to like it. I was lucky to have a father like Dad, who was willing to do whatever it took to give me the education he had wanted for himself, the life, the career that he would have chosen if he could have done it all over.

Each beautiful brick building had a dorm mother, which seemed like an endearing term for housekeeper, and ours was Robin. She was thick around the waist and a heavy smoker; she picked up our forgotten teacups and vacuumed the floors, but more important, she listened to us. We spent mornings spilling the details of our lives to her and to each other.

That was how I came to repair my relationship with Dad when I left home. I reveled in his stories. Recreated my image of him by retelling his stories, like tribal folklore: his world travels, how he'd befriended militias, gotten beat up as a kid, refused to wear a suit or answer the phone or go into private practice. I took pride in all Dad's quirks, shared them with my friends. I remembered what I wanted and forgot the rest. My friends were quick to declare their parents boring and dangerously normal. Dad became a character of fascination. I talked about him incessantly, with admiration.

There were special voices I'd adopted for Mom and Dad, voices that crippled my friends with laughter. "Do your Dad again," they'd say. "Do your Dad ordering food."

And in the voice I'd carved out for him, I'd launch into a reenactment of breakfasts at IHOP.

"No fish in the omelet? No chicken? No beef, ham, or pork? No live animals at all?"

There were pictures I liked to show. "These were their hippie days," I explained. "This was right around the time of Martin Luther King's march on Washington. They kind of raised me going to rallies."

I was the girl with the funny stories. Stories I performed regularly to an audience at dinner, or late night sitting around the living room. I had all of them laughing, making requests to hear certain favorites again. Like the time Dad forgot his passport in Spain and the whole plane waited as a taxi delivered it. Like the time Dad got into a fight with the tourists smoking on the bus.

I rebuilt us like a machine, utilizing the parts that worked, disposing of the broken. Accepting the things I admired in Dad, an admiration that required nothing but my imagination.

I used to bombard Dad with questions that he only half answered, as if there was a side of his life too dark for us to grasp. "Dad," I'd ask. "Did you get drafted for Vietnam?" And instead of answering, Dad would allow a good amount of time to pass before repeating the question.

"Did I get drafted?" he'd say and chuckle, and then again. "They want to know if I got drafted for Vietnam."

From the way he handled questions like this, I could never tell if I'd infringed on territory too painful for Dad to recall or if he simply had avoided the draft and felt guilty for doing so. Often he'd answer a question like this through the telling of what seemed like an irrelevant story, or at least a story not pertaining to the question. Something like, "Did I ever tell you what I saw in the rivers of the little fishing village I worked at in India?"

Dad draped his history in a veil of ambiguity, offering up only occasional clues. Sometimes he'd begin a story and then fade out in the middle, concluding with a rather abrupt statement, like, "Wait until you get out in the world." This was followed by a heavy silence before he'd say, "I mean *really* get out there." And it was the emphasis he placed on *really* that became the foundation on which I built the many imaginary lives I assumed Dad had lived, about which we'd never know.

I had no idea about the actual circumstances surrounding his childhood. Just that we had more than he did growing up. I never knew if they were poor or middle class, though on car rides Dad would sometimes point out various apartment buildings in which he lived as a kid, as well as the schools where he got beat up. Also, we'd visited the Shalom Memorial cemetery, where his father was buried.

I remember asking my mom about Grandpa Joe.

"I don't know," she said. "I don't really know much about Joe."

"You never talked to Dad about him?" I asked, wondering if in Mom's eyes, Dad was born the day she met him. Didn't she wonder what made Dad the man he was? Didn't she question what brought on all his hurting?

When I look at photographs of Dad at his prom and on his bike and in the very first apartment he shared with Mom, I think maybe he was just too young. That Dad hadn't finished growing up himself before I came along. That he didn't know how to relate to children, so that when we demanded to be children he lost all sense of what to do. Maybe Dad had never seen me as a child, or maybe he had and wanted me to remain one forever. Or maybe it never sunk in that I was meant to become someone, that in the same way God had created man, not identical to God but in God's image, Dad had created me, so that I could create myself. I know that in raising us the way he did, Dad saw himself as a model for right. How badly he wanted us to arrive at our destination without straying too much from his path.

I spent years poring over pictures of his youth, examining the expression on his face, the boyish joy in his eyes, or the playfulness between him and his brother, Arthur, on their bicycles, goofing around in the streets of Chicago. At some point, I decided that Dad's childhood had been neither heartbreaking nor glorious, but somewhere in the middle, something closer to normal. And if there was one thing I understood about Dad it was his disinterest in appearing normal. Dad wanted to appear different, superior, at least in the eyes of others.

And yet normality was our shield. We celebrated Shabbat, traveled often and to interesting places, spent the summers away at camp. These rituals encased us in a certain realm of normalcy that allowed us to keep on performing. It allowed me, Jenny, and Mom to cover up what was going on with Dad. For the admiration of outsiders we strived to be seen the way Dad wanted us to appear: better than normal because we were not typical; well-traveled and well-educated but not in the boring, traditional ways. We were normal enough to "pass" among his peers, but superior because we chose youth hostels over hotels, because Dad worked for the VA instead of private practice, because we were the children of the man who had hitchhiked through India and Africa, who refused a frivolous lifestyle in order to teach his children good values.

Dad made it clear that Jenny and I had more than he had ever dreamed. He'd worked long hours as a teenager and got beat up for being Jewish. Grandma Dot always laughed when I asked if they were poor. "Is that what your father says? He's crazy."

So I asked her again over dinner, all of us seated at the table.

Grandma pursed her lips. "C'mon, Steve. We weren't poor."

Dad looked up at the ceiling, then down at his plate of food.

"Grandma, did Dad work day and night at the drugstore?" Jenny asked.

"Not day and night," Grandma said. "He worked a lot when Grandpa Joe was sick."

Grandpa Joe had left Byelorussia at sixteen. Uncle Arthur said he swam across the river between Poland and Germany. In Germany, Joe changed his name to Sontag, which was his stepmother's maiden name. He took a boat over to the U.S. and landed in Baltimore. He became a math teacher. His first wife was Latvian. He didn't like children smacking gum like cows, or lipstick on women. He was a Communist.

"He wasn't a Communist," Grandma said. "He had socialist ideas."

And again Dad would get that mildly amused look in his eyes, as if to say, "Of course he was a Communist."

Dad called Joe Daddy. And perhaps because he'd died before Jenny and I were born, because we'd never touched his skin or smelled his age or shaken his hand (Dad said he had a firm shake), perhaps that was why it made my stomach churn to hear Dad call him Daddy, like a little kid.

"Steve, don't say we were poor," Grandma said, dabbing at the corners of her mouth, "Joe made fine money teaching."

"But we were," Dad said, in a voice that was childishly antagonizing but playful.

Grandma took a second helping of lasagna.

"See, we didn't have second helpings," he said.

"We went to the butcher every week. Tell them about the butcher," said Grandma.

"The butcher liked Grandma," Dad said.

"Oh Steve." Grandma swatted at the air, as if there was something in it she wanted to get rid of. "The butcher cut us big slabs of . . . If we were so poor, Steve, explain the pastrami."

Dad didn't bother to explain the pastrami. He never explained anything. Dad was only interested in presenting the exciting, sometimes unbelievable adventures of an extraordinary Midwestern boy doing the opposite of what society expected.

Later Mom would tell me about her first date with Dad. How he sat her down and showed her all the pictures of the places that he'd traveled to.

"And?" I asked.

"And I was amazed," she said.

I would ask the daughter of our closest family friends, ten

years later, when both of us were in our thirties, how our family had appeared to her back then.

And she would say, "It's the visual, really, that's so overblown in my head."

I asked her what she saw, and she said, "Those smiles. Your huge crazy smiles, that's what I remember. Like everything was so very fantastic, worlds beyond, like this fantastic that only you guys got to have."

She was quiet for a moment, then she said, "I was so scared of your dad. I just never understood those smiles, like I never understood how to get through to you. There were times when I knew something was wrong, but you were so flattened out and shut down, I didn't know how to touch you. You wouldn't make eye contact. God, I hated that! You seemed quite pulled in by your father."

"I was," I said.

"Plus, you were put on a pedestal."

"I was?" I asked, both flattered and surprised. "By whom?"

"My parents," Sarah said. "'Isn't Rachel pretty, and funny and smart and social and adventurous?' You were totally set up as my role model."

"Really?" I asked.

"Yeah," Sarah said. "But there was also that 'See how much Stevie loves Rachel' aspect to it all. I found it very weird that that was so important."

What else did she notice about my dad?

"Incredibly smart, that's what my parents always said, but I just remember him off to the side. In the room, but somewhere else. Actually," Sarah continued, "that's kind of how I remember you, too. Both you and your dad would sneak away from the table and go read something in the den, sitting nowhere close to one another, on opposite sides of a couch or a room, never even acknowledging each other's presence. Both of you were pretty hard to reach."

Then Sarah remembered a time when we were around twelve and she was in my bedroom. NPR was on the radio, and Sarah had asked if we could turn down the volume. She said I told her not to ask questions. She said I was whispering, telling her there were certain things that we just had to do.

"I always felt like I had to perform in front of him. Like everyone changed when he was around. I could see it in my parents, even they changed around him."

But performers stop performing when the show is over. At the end of the day, home was Dad's retreat, the place where he took a break from the pretending. It was the quiet area behind the stage where no one could see or hear us, the place I grew most terrified of.

Our external normalness preserved what we wanted to be known to others and ourselves. Our family history became the safe, reliable stories we could memorialize. Stories I retold like folklore after I left home, because it was the memory of Dad, not Dad himself, which became manageable while I was away at Smith. An intimacy I could control through geographical distance.

■

Sophomore year I declared myself a political science major and joined the cross-country team. I struggled to fit in, to become what I thought a girl at Smith should become. I joined clubs, organized blood drives, canvassed for political groups. But in the mornings I awoke with a claustrophobic uncertainty. I'd overloaded myself with upper-level courses I wasn't prepared for, moved myself to the notorious lesbian dorm so I could make sure my heterosexuality wasn't merely a social condition, isolated myself from my friends back home whose experiences I'd decided were shallow and typical. But all I wanted to do was wade in what was shallow and typical.

I'd met Dave the previous summer in Wisconsin where we worked as camp counselors. He was a junior at the University of North Carolina. He listened well and was good with the kids. He knew how to fix a flat and drive stick shift, and he liked me more than I imagined anyone ever could. I was absolutely terrified of locking myself into a relationship with a man but I wanted to love Dave and I had every intention of losing my virginity to someone I wanted to love.

It was homecoming weekend, the first football game I'd been to. We sat in the stadium at Chapel Hill, watching big uniformed men ram their bodies into each other. Dave held my hand, both of us aware of what was coming, my chest filled with dread and desire, for him and for the act, and for how the act might change me. That evening, Dave wrapped his arms around my neck and lowered himself on top of me. He held my face in his hands, his eyes searched for some indication of my readiness, and it made me think of being in a swimming pool, the seconds in between my last breath of air and plunging under water.

I remember, after the initial discomfort of sex, what a relief it was to feel overpowered by the comfort of another person's body weight. He went to get me water and returned with a piece of fruit. He ran his hands over my belly and covered me up when he saw goose bumps. He asked if I was all right and I said yes and it was nice to just lie there. For the rest of the weekend and the following two years we were together, I relished that feeling of constant surrender. It was such a different form of surrender than what I'd grown accustomed to.

I came to know sex not so much as an act of love but as something that could temporarily quiet me. It was a beautiful, sinking release that shrunk me back to size. Sex was trust, the conscious decision to let a man in, to trust that he'd direct me. How small I felt, and small was how I wanted to see myself and be seen. Later, I got that from other things: running marathons, camping, sky diving, lying on the ground in my sleeping bag surrounded by stars and mountains and the smell of cold air and earth. I grew addicted to my physical insignificance.

Acid had once given me that feeling of insignificance, so I tried it again. Alone, tripping on something I'd bought from a guy at Hampshire College, I wandered around the Smith campus. I was waiting to be infected with laughter, to want to run or swim in the lake or eat an ice cream cone, but I ended up in the stacks at the library, holding open a book, hair spread over my face, crying hysterically. My heart was pounding and my mind was racing and the only thing I felt for sure was the need to hide. I felt huge. I felt like a monster. I felt a million eyes burning into my back, even though no one was there.

When the drug wore off, I sat at Paradise Pond and vomited onto the lawn, more glad than ever to come down. Unlike the other acid trips, this one left me terrified. Not of life or the world or Dad, but of myself. Terrified of the damage I could do to myself if I didn't start making choices. I'd convinced myself that Smith had been my choice, but Smith was Dad's choice, and the more I idealized him, the more I believed my own lies, the more I hated Smith.

I knew the second my feet touched the ground I wouldn't trip again. I also knew I needed to get far away from the person I was becoming. I dropped out of Smith College in the middle of my sophomore year with no idea what I'd do, but certain that something had to change.

SIXTEEN

The bus pulled into Boston's Atlantic Avenue station at a quarter to midnight. It was January 15. I had a duffel full of clothing and a check from Dad for $200, which I was set on returning unused.

Boston was just two hours by bus from Smith, and I had friends at Smith who would take me in if I failed to find my way in Boston. I'd briefly considered internships at television networks in New York, but the thought of the city overwhelmed me and I couldn't have supported myself. Chicago was too close to home. Boston was small and contained, a manageable place to navigate, and with all the students in Boston I thought I'd be able to camouflage my failure.

With no place to sleep that first night I wandered around the small, dilapidated bus station. Rows of heavy plastic chairs filled the place. Black-and-white TVs that cost a couple of quarters to view extended upward from the sides of the chairs. The agent announced that the station would be closing. I hadn't planned on that. I hadn't planned a thing. A red-bearded man kicked the vending machine, his candy bar stuck inside. I got up from the chair and walked toward the station agent, who had a hard Boston accent but soft eyes.

"It's just not something we do," he said, shaking his head at my duffel.

"Could we pretend that I just left it on the bus?"

He looked at the man at the vending machine. "This isn't a hotel," he whispered, "it's a bus station."

I smiled at him, hoping he didn't think I had no place to stay. "It's just my bag that needs a place, not me. What if I'm back here at five in the morning?"

"Wouldn't do much good, we don't open until six."

Hesitantly, he tagged my bags.

"Wouldn't do this for just anyone," he said.

His hands were rough, with patches of small, red freckles around his knuckles. I wrapped a wool scarf around my neck, thanked the man, and stood outside at the bus stop. The cold felt heavy on my face. Awake, I thought. This is how it feels to be awake. I stood there for a good twenty minutes before a cab driver pulled over and told me the buses had stopped running for the night. I asked him to drive me to the closest IHOP, the only twenty-four-hour joint that came to mind. He dropped me off on Commonwealth Avenue. I ordered eggs over easy and a large pot of coffee from a waitress who couldn't have been much over thirty, with yellow teeth and hair to match. She was vacuuming around the booths when I sat down. I took out a pen and paper. The waitress resumed vacuuming in between taking my order and bringing my food, apologizing unnecessarily for the noise.

"You studying?" she asked.

I nodded and smiled. We were the only two people there. She knew Boston, I could see it in her face, probably lived there all her life. I asked her for a cigarette; she gave me two, and something about her giving me two made me think I'd find my way in Boston. I thought about the lyrics of Cat Stevens's "Wild World," thought about it being hard to get by just upon a smile, and about Dad warning me to watch whom I smile at. I thought of the waitress returning home, sinking into her sofa, taking off her shoes one at a time, lighting a cigarette in some dimly lit apartment.

The waitress dropped more cigarettes on my table as night turned into morning. She brewed another pot of coffee. I thought if I stuck around long enough that she would take me back to that dimly lit apartment, spread an afghan over me and let me stay. I could tell she was lonely and might have liked my company. I relied on people liking me.

"All-nighter?" she asked.

"Until I finish this paper," I said.

I went in and out of quick, hard sleeps, my face on the table cradled in my arms. When it was light, I went back to the Greyhound station for my duffel and then I stuffed all my clothes into several small lockers, each costing a quarter. I got a token for the bathroom, where I brushed my teeth and washed my face. The station agent, no longer the soft man from the night before but an angular-faced red-haired woman, pointed me in the direction of the tourist center at Faneuil Hall.

I moved through groups of schoolchildren on field trips, and tourists slipping on ice. I walked into clothing boutiques, disappeared into dressing rooms and tried on hats and coats trimmed with fur and lace, which made me look like a stranger to myself. At the tourist center, I took free maps of the city, a listing of youth hostels, several books of coupons. The pubs were filling up with men. I envied the daytime drunks and wished that I were old enough to pass the day in a warm, dark place loud with music and laughter. I passed chess players in the park and fat flocks of pigeons in the Public Gardens. I cut across the city, eventually finding my way down to Boylston, a street full of record stores and ice cream shops and stringy musicians. Then, counting the remaining daylight hours, I tried to think of whom I knew in Boston.

By dark, I was back at IHOP. A different waitress came around with coffee, not the kind of waitress I would ask for cigarettes. There were families eating dinner, kids drinking Pepsi out of big plastic cups, forking into bacon strips and slabs of French toast. A man and a woman sat in a booth near mine; they were young, probably just out of college. I assumed they knew each other well—or not at all, from the way they watched their plates and not each other.

Dad's reaction to my dropping out of Smith surprised me. I'd prepared myself for the worst. I was ready to plead to be sent to another school. I'd talked to the dean of students, gotten all the information I needed to transfer, and was ready to battle any and all of Dad's concerns by telling him Smith was a make-believe land and I wanted hands-on experience in the real world.

In fact, that was all I had to say. Dad seemed more interested in what I planned to do than disappointed by my quitting Smith. It was Mom who asked the questions. How would I manage? Where would I live? Had I gone to the career counselor at Smith? Did they have any kind of internship program through the school?

And because it was Mom asking, I didn't think I had to answer.

Dad was only half in the conversation when I said, "So, you guys support me on this?"—meaning there would be no delayed repercussions down the road, no surprise punches. We were all in the living room. Dad was setting up the VCR to tape a documentary on PBS.

"It's *how* you spend the time, Rachel. That's what's important to me."

"So you do support the decision?" I asked.

Dad stood up from the VCR, walked over to his desk, and said, "Absolutely."

He looked at me, nodded, then turned to Mom and said, "Ellen, do we support Rachel's decision?"

That conversation took place over Thanksgiving. I'd already filed for a leave of absence and withdrawn myself from the spring semester and I'd gone home expecting to be disowned or cut off. I was met with not only support but something that felt like love. I felt it when he handed me the check, I felt it when he dropped me at the airport, the way he said good luck and let us know what happens, in the same way he liked to say, "take an aspirin call me in the morning."

And then when I got on the plane I feared something else. I feared Dad's certainty that he'd get my phone call in a couple of weeks, admitting that I couldn't make it on my own.

From the pay phone outside IHOP I got up the guts to call Emerson College student information, where I asked for the number of a boy I'd kissed years before on our trip to Cancún. His name was John and I was lucky to have remembered his last name. It took him some time to remember me, and when he did his voice turned odd, the way orange juice tastes after brushing your teeth.

Twice the pay phone operator requested more dimes before I got an address out of him. Twice I considered hanging up and calling someone else, but it was dark and I wanted to get somewhere. I hopped on the T to his dorm, where he played me recordings of the music he made, and where he let me sleep on his foldout futon for several nights before I remembered Jack.

Jack was a big, solid redhead from Arkansas and a fellow congressional page. Someone who'd make a fine senator, someone who never went unnoticed. I stood at a pay phone in Cambridge, trying to sound casual.

He repeated my name several times in his slow Southern drawl, and I knew by the hearty laugh he gave when he was done repeating my name that he would let me stay with him. I told him I was taking time off. I told him I was living in Boston.

"You should come on by sometime," he said.

His Harvard suite was small and smelled like boy. The walls were plastered with pictures of baseball players and supermodels, the floors were littered with books and beer stains. Handsome brown tables decorated the lobby of the dorm, an old converted eating club. Every morning, I left the suite when Jack left for class. I walked the city, looking for housing, and returned in the evening to drain pitchers of Michelob at the Crimson Bar, where we inevitably ended up belting out Harvard cheers I didn't know the words to.

Jack's friend took a liking to me. He was a sharp-boned baseball player from Connecticut, the kind of guy who said, "Chivalry's not dead" while holding open the door. I didn't like him much, but accepted half of his bed after spending several nights on Jack's couch. It was late and we were drunk. The friend threw me a T-shirt and promised not to try anything. I didn't really care much if he did. I'd already managed to keep myself at Harvard for a week.

I woke up to him masturbating against my leg, his reckless,

boyish breath grazing my shoulder. I closed my eyes and let him finish, turning my body toward the wall. I placed his arms around my waist and tried to fall asleep.

It wasn't entirely that I had to prove I could make it, but rather I wanted to expose myself to danger and survive. I wanted to test my instincts, my resourcefulness, see just how far I could get on nothing. I wanted to know that regardless of what I came up against in life, Dad was the worst thing I'd have to go through. That nothing would ever be that bad again. And the only way I could prove that was to put myself in situations I could not control or predict.

The youth hostel on Commonwealth Avenue cost $25 a night. That gave me three nights without cashing Dad's check or ten if I caved. The woman at the desk waived the nightly fees and let me work instead. I pulled hair out of drains and stripped linens from beds. The hostel closed for six hours during the day, time I used to roam around Boston looking for work and a place to live. I collected my clothes from the quarter lockers and brought them to the hostel. I called Jack to thank him for the hospitality. Then I called my parents and left a message on the machine.

"I had an interview," I said. "It went really well."

My lies came easily. I was hardly even conscious of them. They felt like nothing more than confident predictions. I would have an interview. It would go well.

I applied to be a waitress at Jillian's Pool Hall on Lansdowne Street. I applied to check coats at Axis Nightclub, to bus tables at Cask n' Flagon. I dropped résumés at bookstores and bars, always writing the address of the youth hostel in place of the address I didn't have. What I really wanted was a job I could call home about, a job with John Kerry or Ted Kennedy. Morning coffee from the Au Bon Pain before I headed up to the office, happy hour invitations, wry exchanges with other interns.

I'd been eating my way through various college open houses. If I ate enough cheese in one night, I could keep myself full until three the next day. Breakfast consisted of coffee, three or four cups at cafés where refills were free, and half-priced day-old bagels from Store 24. Dinner alternated between a pizza slice and the coffee/muffin combination at Dunkin Donuts for $1.25.

There were tons of free art openings listed in *The Phoenix*, Boston's alternative newspaper, and I went to every one that had a reception where I could stack my plate with cheese and crackers, fruit and cake, cookies, vegetables, breads. Wednesday nights at the Boston Museum of Fine Art were free, and so were Thursdays after five at the Institute of Contemporary Art. The Historical Museum was always free, as well as the Harvard Botanical Garden, MIT, the Old North Church, and the USS *Constitution* Museum. I took long walks through the North End, where I was often mistaken for an Italian and given free cans of Coke with my pizza. Universities organized fully catered receptions after lectures, and I sat listening to Noam Chomsky, Elie Wiesel, and Cornel West. Usually I could account for both lunch and dinner at Simmons College, where their student art openings laid the most elaborate spreads. I was afraid of becoming dumb since I'd left school, in the same way I was afraid of becoming weak since I'd left Dad. I found bookstores all over the city, where authors read from their new novels. Sometimes I listened and sometimes I just sat in the back waiting for the food

The student center at BU was a great place to loiter. It was full of everything I needed. Bathrooms, cheap cups of coffee, rows of ten-cent pay phones, bulletin boards crammed with sublet signs. I wrote down several telephone numbers of students looking for roommates.

The first girls I called had a room for rent at $250 a month. They were students at Boston College, looking for someone "nice, independent, and clean." Both girls were in slippers when I arrived. The heater was on high and their place exuded an uncomfortable warmth.

We sat on the couch. They offered me tea. I felt like it was something I was supposed to refuse, but I accepted, hoping they'd bring food, too.

The taller of the roommates brought my tea. "Rachel's a name from the Old Testament," she said, placing it down on a coaster.

"It means ewe," I said. "Like lamb."

"Are you religious?" asked the other.

"Kind of," I said.

"Both of us are as well."

The girls' smiles grew achingly big and for the first time since I'd arrived in Boston two weeks earlier, everything came to a standstill. I wanted nothing more than to wake up in the same place every morning. One of the girls leaned in when I spoke, listening with her entire body, as we talked about her just-married sister, how marriage was the ultimate, how lucky she'd been to find a man who believed like she did.

"Have you been religious all your life, Rachel?"

"I went to Sunday school," I said, liking the role of prospective Christian roommate.

"It's important for us to live with someone who has similar values," said the tea bringer.

"Definitely," I agreed.

"There are so many people who just don't seem to care."

It was an afternoon of conversational vagueness and blasting heat. It was a reminder that I could adapt to any situation. It was the greatest gift Dad gave me. He'd shown us worlds outside of our own and exposed us to different cultures, which developed a confidence in me that I would use like a passport, moving through the world as I pleased.

"So many people have lost faith in miracles," said one of the girls.

Again I agreed, but I was no longer sure to what.

Their smiles were becoming sickening, almost revolting to me. Was revolting too harsh? Was I too judgmental? What the hell was I doing? There was a brown leather-bound copy of the New Testament on the table—I had seen it immediately upon sitting down, and I sat down anyway. I sat down because I wanted to hear their stories. At what point did they decide to go along with God? Was it a birthright that they never questioned or had they found God at Boston College? I sat down to find out what made them who they were, to see how well I could become exactly who they wanted me to be. I sat down because I needed a place to live. They needed a security deposit. The gratification I'd get from returning Dad's two hundred dollar check was well worth continuing the search for a place to live.

Before I left, I used the bathroom. It contained a bathmat that read TRUST. I sat on the toilet with my hands over my face, thinking again about the heavy breath of Jack's roommate before he came on my leg, and how I hadn't even bothered to wash it off.

Back out in the frigid-cold Boston afternoon, the sky was a dull gray; it could have been any hour. In an alley around the corner from the girls, I slowly began to convulse into tears.

Although it occurred to me several times that I didn't have to do what I was doing in Boston, I didn't know what else to do with myself.

I'd convinced myself that Boston was an adventure, a test of character, freedom. It was the equivalent of other kids' first year at college, an experiment in living away from the comforts of home. But my experiment was more about how far I'd have to go in order to be accepted or disowned by Dad. And because I knew Dad would never accept me, I needed him to reject me, to tell me I could not come home. I was waiting for his permission to move on.

I didn't know how to take care of myself, only how to protect myself, so I went around looking for and finding discomfort, seeking free ways to keep warm, like sitting at the airport, and watching planes take off. But it wasn't warmth I wanted; it was the continued reassurance that I could survive anything that came my way.

I got in the habit of creating worst-case scenarios. Visualizing my reactions detail by detail.

How would I handle getting lost on a hike, in varying temperatures, without food or water? What would I use for warmth and shelter? How would I deal with the death of a husband? Or the death of my own child? How would I handle getting mugged or, worse yet, raped? It seemed imperative that I had some sort of understanding of how I would deal with a physical threat, because I certainly knew better than to call home if something like that happened.

I walked myself through the options. I could talk to the rapist, get him to open up to me, get him in an emotionally vulnerable place so that I could make my escape. If that was not possible, I'd succumb, demand that the rapist use a condom, go to that place in my head where I could shut down until it was over. And when it was over, I would calmly wait for him to leave, showing absolutely no emotion, then I would calmly call the police. I would remain calm while they questioned me in my apartment or down at the station or on the way to the hospital. And then I would call a friend and calmly instruct her to bring me fresh clothes and take me home. I even had the phone call rehearsed on both ends, how I wouldn't mention anything about the rape until the friend arrived. The important thing was that I was still alive and, loving life as much as I did, would soon recover. But what was really important was the impact I would make on those police. How impressed they'd be with the way I handled both the criminal during the act and myself after the fact.

It was an arrogant way to think, but I felt so equipped to save myself from anything and simultaneously terrified of not being fully prepared for what could happen.

I started work as a cocktail waitress at Boston Billiards, an upscale pool hall on Lansdowne Street. By the end of the first week, I'd been demoted to the desk, where I rented tables and upgraded pool sticks. My shifts began at six and ended at two. I drank coffee all night, collected bribes from big shots who slid me twenties to knock them to the front of the wait list. At the end of the night, I slipped out alone and walked back to the hostel. I didn't want to spend my tips sharing cabs.

I left more messages on the answering machines of prospective roommates, messages that could not be returned because I had no number. I tried my way into the student housing offices at Boston College, Emerson, and Northeastern. At each I was gently informed that the facilities were for students of the university. I sat in various food courts waiting for people to sit down and talk to me, thinking if I could find myself a student ID for a day I could find that place to live. Walking killed the days, and I walked the entire freezing city.

In the lobby bathroom of the Boston Sheridan, I ran hot water over my arms and all the way up to my elbows. Then I took paper towels, soaked them in hot water, and warmed my neck. My face was red from the cold, my lips were bloody from cracking, and a layer of snot had dried around my nose. Still, the mirror reflected a girl in a mauve Anne Klein scarf and a gray wool hat. I looked normal, somewhat presentable.

The pay phone booths were handsome and made of dark brown wood, each equipped with a seat. I closed the door and dialed home.

"Do you know that you have a cousin in Boston?" Mom asked.

Carrie was Mom's second cousin and had graduated years ago from Mt. Holyoke. Mom thought she might be a curator at a museum, a nice young woman who'd get a kick out of hearing from me, although we'd never met. I asked to speak to Dad.

"Look who it is . . . look who it is!" he said.

I pictured him in his study with his headset on, his computer screen glowing. "Rachel, what's the good word?"

I took a long breath. "Well. Here I am."

"And how's Boston?" Dad asked.

"It's fine."

"Okay," he said. "Okay."

"It's cold."

"Okay."

"It's challenging, Dad."

"How?"

"I haven't found a place to live, you know, so it's kind of forcing me to open up my eyes to all I have. I'm realizing, I mean, I've always known, but it's really sinking in right now, how lucky I've been, how privileged. I mean, it's cold out here, you know? It's cold and I'm without the usual comforts . . . Dad?"

"I'm listening," he said.

"But at least I have the resources to find my way around, to get myself food. I have the social skills to work my way into a bed at the youth hostel and the confidence that it'll pass. Just knowing that you guys are there, if I fail. Not that I'd expect you to pick up my slack but still it's different for me, being middle-class and educated. You know. It will always be easier."

"Yup."

"But it's ironic, too. I feel more affected by things. I bought a bag of day-old bagels at Store 24 the other day, and I'm walking down the street and I see this guy, this homeless guy with a Styrofoam cup asking for money. I look him in the eye and say, 'Sorry, I can't help you out with money, but do you want a bagel?' Dad . . . ?"

"I'm listening."

"So I give him one and he takes a bite, I had already eaten one and they were fine, and he throws the bagel back at me and says, 'Don't give me no stale bagel.' It was amazing, Dad. I was mad. I don't even know why I was so mad, but it was pretty insulting, him throwing the bagel back."

"And?"

"And, I'm learning. Glad I'm doing this, Dad. Not sure exactly what I'm learning, but I am."

"You know, when I was in Africa—" he began.

I watched people get on and off elevators, women with shopping bags, flocks of corporate conference-goers. I tried to imagine myself older, standing in an elevator, briefcase in hand, maybe as a lawyer. I could be a lawyer.

"—I was hitchhiking in the back of a military supplies truck, with these guerrillas who wanted to kill me."

"Right," I said.

"There was no reason for them not to kill me and I knew this."

"Okay."

"And I slept wherever I could sleep. But I never really slept, because my life was on the line. But when I did, I'd just throw myself down in a bush or in the back of a truck, in some village somewhere. I had no idea where I was going, where I'd end up at the end of a day. I was completely at their mercy."

"Dad, I was freezing the other day, literally, my lips were blue and I was shaking and I walked myself into this church and just sat there for hours. Just to keep warm. I'm staring up at this huge stained-glass vision of the Virgin Mary and there's an enormous crucifix planted on the pulpit, which would normally make me feel a little shaken, or eerie, or something, but that place was nothing other than warmth on a really bitter cold day. The days feel, like, pretty unreal in a lot of ways."

"They feel, like, pretty unreal? Or they feel unreal?"

"They feel unreal."

"Really, Rachel, here you are, nineteen years old, in the middle of saying something interesting, and then you have to ruin it with like. Like. Like. That's all I can hear."

I began feeling claustrophobic in the phone booth. My head was flooded with images of hands. The station agent as he tagged my bags, the IHOP waitress dropping cigarettes on the table, Jack's friend's hands locked around my waist.

"Dad," I said, "I get it."

"It's like when you had that nose ring and you were sitting at Pizza Hut talking about a political action group and I couldn't hear a word you said. All I could see was that rod in your nose. You get it?"

"I get it."

What I got was the urge to hang up. What I got was a sense that Dad had drained himself of any kind of parental concern when it came to me. I thought about the pepper spray key chains he'd bought for us years ago, and how, every time we left the house, he'd ask if we had our canisters. Mom had accidentally sprayed herself and gone to the hospital, and I knew it didn't make much sense carrying around something that could be used against me, but I knew it had come as a gift from that part of Dad that hoarded love like a chipmunk does acorns. He felt an enormous burden to protect his family from the world, and for a long time I carried that pepper spray around.

"Well, it sounds like things are going well," he said. "It sounds to me like you're learning."

This was the highest compliment he could pay, the silent acknowledgment that not only was I learning but that I, like him, was dedicated to doing it on my own.

"Some days I feel like I'm really going crazy."

"Yup," he said, "I know how that feels."

What I really didn't get was why he could say yup but I couldn't say like.

"Well I'm going to let you go," he said. Something in his voice began to soften. It seemed a funny thing to say, since I'd called to talk to him.

■

Ten years later, talking with my uncle Arthur and my cousin Jill, I learned Dad had been proud of my time in Boston.

"Your dad bragged about your homelessness like a parent gloating over their child's college scholarship," Jill said, recounting a night when Mom and Dad had gone to their house for dinner while I was in Boston.

"Rachel's on the streets," he'd said. "Wandering around on the subways, sleeping in churches. She called the other day. We don't have a way to get in touch with her. Do we, Ellen?"

"You're happy that your daughter has no place to sleep?" Jill asked.

Dad looked shocked. For a quick second, he looked down at the table, which Mom had been staring at all night, then he looked up and said, "Rachel's making it on her own. She's working hard to make it on her own. Don't you get it, Jill? Rachel's gonna make it."

■

On a brown leather chair in the lobby of the Sheridan, I wondered what it might take to get a free room, or at least convince a maid to let me take a bath. I flipped through a magazine of Boston's attractions and thought half-heartedly of getting a credit card.

I hadn't told Dad about the kindness. The youth hostel clerk who'd given me work, the waitress at IHOP, the stranger who'd picked up my hat from the street when it dropped from my pocket. I lived for that kindness, the passing anonymous exchange with strangers. The momentary, fleeting feeling of being understood. It terrified me, the things I would do for that approval.

■

I looked at four more apartments. Two required security deposits, a term I'd never heard before.

The third apartment was a share in Allston with a woman in her fifties, who had hamsters running freely in her bedroom. For two hundred a month, I could sleep in a large walk-in closet. No one answered the door at the fourth apartment.

Monday morning I left the hostel and set out toward Federal Center, where Senators Kennedy and Kerry had offices.

The receptionist in Kerry's office was an older woman in a soft knit sweater who, after learning that I'd dropped out of college, said, "Send us your résumé."

I placed it in front of her. "Can I leave it with you?"

"We ask that people apply through the mail."

"Any exceptions?" I asked. "Because I really want to work."

"The person that you'll want to speak with is David DeMartino," she said. "He's the one that makes the decision for this type of thing."

DeMartino was a jovial man, with Michael Dukakis–thick eyebrows and a crushing Boston accent. He was most impressed with my congressional experience, since he, too, had been a page "back in the day." We sat in his office, trading stories, every now and again referring to something on my résumé—time spent as a candy striper at St. Francis Hospital, the hostess job at Bennigan's.

"Well, we can't pay you anything, since you don't have a degree," he said, "but we can definitely make good use of you."

Boston had been given money as part of President Clinton's Empowerment Zone Fund, and my boss was responsible for deciding which neighborhoods and organizations would get what grant. Within half an hour I'd been assigned to work in the office of Low Income Housing and Social Welfare, where I'd spend most of my time responding to letters of complaints from constituents, most of them elderly people with too much time on their hands.

I sat down in an Italian eatery in the North End, ordered a cannoli, and rehearsed the phone conversation I planned to have with Dad. The David DeMartino connection might demean my merit, make me seem more lucky than resourceful. And walking into Senator Kerry's office with no game plan sounded lazy. It made the most sense to say I'd applied months before. It made the job sound more prestigious, competitive, my position legitimate and deserved. I wouldn't tell Dad that I made my money waiting tables at a pub and renting pool cues at Boston Billiards.

We were, I thought, engaged in some sort of unspoken game. Don't ask. Don't tell. Drop out of Smith. Go find your own way. No punishment. No reward.

In some ways we had a competition going. I didn't see it then but I was rejecting what Dad saw as a very good offer. He was willing to pay my way if I was willing to become the person he wanted me to be. Small deviations like Boston would be allowed if I got back on track. I wanted to make Dad proud but I also wanted him to acknowledge that I didn't need him. I wanted him to think my life had become important after Boston; that I was content and his reactions no longer made the impact they once had. I used to imagine the casual way in which I'd tell Dad about my job with Senator Kerry's office.

"One of my more interesting gigs in Boston," I would say. And I'd change it to Kennedy. Kennedy sounded better.

But what I really wanted was to have Dad be interested in me. I wanted him to ask how I'd managed. I wanted him to ask if I was scared, if I was lonely.

I was both. Sitting in the cafés, the hotel lobbies, the student centers. The feeling of defeat that came at the end of each day, a feeling I'd based entirely around the notion of making it or not making it. I still had no place to live.

I considered calling Jack again. I considered going home. Instead, I dialed Cousin Carrie. She sounded proper and a little reserved, a part of the family we'd never met and rarely spoke with. Her father was Mom's uncle. Carrie invited me over right then and there, and we sat in her tiny studio apartment on Beacon Street, drinking scotch and sodas.

"It was one of the very first things I learned how to do," Carrie said, laughing, "when my father returned from work each day."

I tried to imagine Dad drinking scotch. Or better yet, to imagine Jenny and me running toward a mirrored tray to fix him a drink, with Mom ever so slightly in the background.

Carrie was tall, with fair skin and short hair. She wore cashmere sweaters and Alfred Sung perfume. She'd been taught to date many and marry late, to pursue a distinguished career. After a few more scotch and sodas, Carrie invited me to spend the night, a good thing, since I had no other plan. She lent me flannel pajamas, gave me a toothbrush, a bar of mild French soap. We sat on the floor, smoking her stale Benson and Hedges, waiting for grapefruit masks to harden on our faces.

"Do you smoke a lot?" I asked.

"No, actually, I keep this pack on reserve for special occasions."

Flattered, I wondered how long special occasions could stay before overstepping their welcome.

"What do your parents think about you dropping out of Smith?" she asked.

"They're supportive."

"Really?"

"My dad's a real free spirit," I said. "Very open-minded about things like this."

"That's lucky. My father would have been bothered, I think."

"My dad's seen a lot of life."

"I don't know your side of the family at all," she said. "But my father loves your mom, and I hear your father has quite the sense of humor."

"Oh, he's funny," I said.

I called home from Carrie's telephone. Mom was delighted I was reaching out, although my reaching out had become more like a polite form of begging. Carrie took great care of me. I stayed for a week, leaving when she went to work.

I found my home on a flyer, taped to a streetlamp outside of T. Anthony's Pizza.

ONE ROOM IN APARTMENT WITH THREE OTHERS. NO LIVING ROOM. NO SECURITY DEPOSIT. $250. SUBLET THROUGH MAY. CALL WILL.

The apartment was a mile from Boston University. Will and his girlfriend shared a room off the kitchen, Dwayne had his own, and one was unoccupied and for me, if I wanted it.

"Can you take a look on Tuesday?" Will asked. He was at his folks' place in South Boston for the weekend.

"How about today?" I asked. "I could move in today."

"Dwayne might be home later on."

"Well, if it's okay with Dwayne, I'll move in tonight."

■

A large wooden bookshelf and a futon had been left in the bedroom. I unzipped my sleeping bag and lay it across the futon. Out my window I stared at the backs of brick apartment buildings lined with snow-covered fire escapes: six huge garbage containers down below, a scattering of naked trees, illegally parked cars.

Mine was the only room with a real door. I loved that room, the fact that I could shut the door and no one would disturb me. The four of us shared a phone that sat on the floor in the hall and I loved to hear the others answer when it was for me. How they'd suggest I might be sleeping, but they weren't sure, and should they knock or take a message. That simple thing left me feeling so protected, which was so much of what I needed that cold, gray winter.

I cried a lot. Not always for any distinguishable reason but because it felt safe, and I had the privacy I'd always craved. For the first time since D.C. everything inside me stopped moving and I felt like I could breathe.

My roommates became an integral part of my time there. Especially Will's girlfriend, Wendy, who'd just returned from a mental institution after attempting suicide. We would borrow her mom's car and drive up to Walden Pond when the weather got nice, lie on the grass in our sweaters, and read each other parts of our favorite books. We would drive for hours with no place to go, drink beer, paint, and play guitar in the kitchen.

I had very little interest in talking to Dad once I found home. Mom called on occasion. Sometimes I picked up, but for the most part I kept our conversations limited to practical matters, like where I'd be going next.

I'd applied to study abroad for a year, in Jerusalem. It was a cultural experience I wanted for myself, and one for which I had Dad's blessings and tuition payments. I mistakenly thought that my going to Boston had given Dad a deeper understanding of me, as going to Israel would do. My mistake was in still wanting to be understood. As much as I tried to prove I could make it without Dad, he loomed over everything I did. My internship with Kerry was over, and I was ready to move on.

I was attracted to having a home, a place I could return to, and at the same time I sought adventure, newness, travel. If there was anything beautiful I'd inherited from Dad it was my wanderlust.

In May, when the sublet ended, so did my time in Boston. I was convinced I could make a home out of anything. My goal in traveling was to find a deeper sense of comfort than what I'd found anywhere so far. Boston began a decade of living and leaving. I liked the idea of arriving with nothing and finding something, whether it was a job or a home or a lover.

From cities to relationships, I hunted comfort until I found it and left before it got too easy. Even when I was physically present I was constantly leaving. It was what I did best.

Soon after Boston I went to Israel for a year abroad. Within a week of my arrival, the first of many buses got blown up in terrorist attacks. This one was on its way to the university, three buses ahead of the one that I was on. Several students were killed, and after hours of sitting on the bus while the paramedics cleaned up the mess of bodies, we were taken the rest of the way to the university. Later we walked across town and lined up at the Red Cross to give blood to the victims. Six months later, Prime Minister Yitzhak Rabin was assassinated, then more bombings. Parents of American students demanded their children come home. Hundreds left.

"We're proud of you," Dad said to me one night on the phone. "And disgusted with the people that are leaving. What cowards."

"It's not their fault," I explained. "It's their parents. Their parents are the ones who are afraid. They don't even know what's going on."

"We wouldn't let you come home if you wanted to."

"I wouldn't want to," I said. "Anyways, Dad, I'm proud that you know better than to ask me to come home."

I tried hard to believe that Dad and I saw eye-to-eye on right and wrong, and that we had been able to pull our relationship together. Dad seemed to have finally recognized me as an adult, and because of this, and my geographical distance, we could finally enjoy some peace. It appeared that even the best of families maintained some type of superficial relationship.

This was how I explained things to my cousins when they came to Israel for a visit. We were sitting at a café in Tel Aviv. Debbie and Jill were looking at me as if I'd just told them I had a disease that wouldn't kill me but would periodically make me suffer for the rest of my life.

"You're proud of your dad?" Jill asked.

I actually remember the earnestness with which I nodded.

"Don't you think he should be proud of you?" she asked.

"Oh it's mutual," I explained. "I mean it's like we've finally found that common ground."

Jill and Debbie looked as if they didn't know how to proceed. Although they never looked at each other, I could tell there was a whole conversation happening over me, some agreement that my disease was more serious than previously diagnosed. And even I could feel myself divided, doubting my own words, regretting the certainty with which I was presenting my newfound relationship with Dad. It was different than the way I described him to friends at Smith. I knew as well as, if not better than my cousins, how sick my father was. My need to be accepted or rejected by him was stronger than ever.

"It's not okay," Jill said. Not once but over and over.

I heard her and I heard myself, trying for one last time to find a way to keep my father in my life while beginning my own, something I now know was absolutely impossible. I thought if I could just find a place to stow away Dad, like a piece of oversized luggage, I could maintain the façade of normalcy.

I let myself believe that everyone else was a coward, that I was strong, that I didn't want parents who'd compromise their political beliefs for their peace of mind, that I didn't wish Dad and Mom would ask me to come home. I lied to myself.

"I'm not angry anymore," I insisted.

Like Mom compromising herself to maintain a relationship with Dad, I could feel myself fluctuating. Knowing I'd never be loved by Dad the way I wanted but thinking, after all the death I'd seen that year, that I owed him one more try.

Twelve years later, Jill and Debbie still refer to Israel as the time they worried about me most. "God, you were far gone," Jill said. "That look in your eyes, how militantly you stated everything was fine at home, we thought for sure we had lost you."

They weren't the only ones.

At twenty-one, when I returned from Israel, I transferred to the University of Wisconsin. I started dating a guy named Keenan, to whom I told everything. Each visit home, every phone conversation, brought on a mess of confusion.

Keenan, who ten years later would remain one of my closest friends, refused after a while to talk about my Dad. "Rachel," he would say. "It kills me to hear. It breaks my fucking heart."

Just before winter break, Keenan asked me to go to Minnesota with him. I'd planned to go home.

"Just come to St. Paul," Keenan said. "Or just stay here at the co-op and I'll teach you how to play guitar." But I was set on going home. I was ready to move on, and I knew whatever it was that would send me over the edge was waiting for me at home.

"I'm going back," I said to Keenan. "But if things go bad, I'll leave for good."

I could hear my Mom inside me. I knew how ridiculous I must have seemed to someone like Keenan, whose views were smart and simple: if it's hurting you, it's no good.

"It's more complicated than that," I heard myself say. "It's family."

But I knew what I would no longer do for my family.

Very decisively, with a stern confidence that defied his usual soft-spoken manner, Keenan touched his hands to my shoulders and said, "Really, Rachel, under no circumstance should you return home."

But I did. I returned for the last time. Dad had been hounding me all semester. I had amounted to nothing. Had I learned anything working for the senator from Massachusetts or paging in Washington? Had I *really* done research or had I just shuffled papers? Did I know what an ignorant, worthless waste of a child I was?

Dad had ideas about my future, what I could do with my life that would make a difference in the world. His world. I could be a journalist or work for the Foreign Service. But by then I'd stopped listening. And it wasn't to block out the noise but to take in with my eyes, for the sake of memory, my family. I knew the end was near, and I cried myself to sleep and I cried openly at the dinner table. The crying felt similar to the emotional exhaustion at the end of a long breakup, when both people, their hurt blurred and their hearts beginning to untangle, can see for a second the person they once loved. Like a sinner who'd tried and failed to be holy, I was ready to be excommunicated.

I'd bought a plane ticket to Colorado to spend a week of winter break visiting my friend Laura. But Dad was set on my going back to Wisconsin and studying for the GREs. He made it clear I was not to get on the plane. My ticket would go unused. I would get on a bus and go back to Madison.

The plane that I was not supposed to catch flew out of O'Hare Airport. The bus back to Madison also left from a terminal at O'Hare. Dad insisted on dropping me off. I thanked him for the ride, and raced to the other terminal to check in to my Denver flight.

Mom had come into my room the night before. "Stop fighting," she said. "Just get on the plane and go to Colorado." She handed me a couple of twenties to take Laura out for lunch. She'd packed me a bag of candy and nuts for the plane. "Dad's not going to know the difference."

We were together in our lies. All of us, that's how we got by. Our lies were our unspoken way of living around the unreasonable.

■

Three months later, on a Saturday in March, I called Mom from a pay phone at the Canterbury Booksellers to find out where she was. She'd planned on visiting me in Madison that weekend.

"There's been a mistake," she said, and from her solemn tone, I was sure she meant a death. A mistake was adding the wrong ingredient to a cake. Throwing a baseball through the neighbor's window.

The mistake turned out to be mine. I'd used Dad's calling card from Denver. I'd been caught. But Mom was willing to fight the fight. She had managed to convince Dad that the charge was a mistake, and that I had, in fact, not gone to Colorado but gotten on the bus back to campus. From the slow, annoyed way she was speaking, I could tell he was beside her. Mom was on trial.

I also knew that if we didn't stop lying, we'd forever be stuck in the maze.

"I'm not going to lie, Mom."

"Rachel, no one's asking you to lie."

Mom no longer seemed able to distinguish a lie from the truth. She saw it as a simple, necessary request. She had created a story which I needed to stand by in order for her to be acquitted.

"Don't do this," Mom whispered into the phone. "I'm telling you it's just not worth it."

But what was worth it for Mom was never going to be worth it for me. She'd spent her whole life maintaining a relationship that suffocated her, and where all the lies and fights seemed to circle around me. I was not a part of their reality anymore, nor was I willing to be pulled back in.

"It's very simple. Just tell Dad you didn't go."

If I told him that I hadn't gone to Colorado, he probably wouldn't have challenged it. Mom would be able to get in the car and drive herself to Madison. All would return to normal. And that was my biggest fear, that I'd go on faking my way through life. I lied all the time without realizing it. Little lies but to people I didn't need to lie to. I told friends I was going to the library when I was going to the movies. I told boyfriends I was sick in bed when I needed time alone. I simply wasn't in the practice of telling the truth.

Even over the phone, I could hear Mom's fear when Dad picked up the line, like her breath was trapped in her lungs.

"Did you go?" he asked.

"Do you remember Dad asking you not to go to Colorado?" Mom asked. Both of them were speaking simultaneously and if there was any part of me that had considered lying just once more, hearing Mom's convoluted wording made me stop.

"Ellen," Dad said. "That's not the way to go about it."

"Steve, I trust that Rachel did not go."

"So the ticket went unused?" he asked.

"The ticket was never used," Mom said.

"I want to hear it from her," he said. "Because Rachel, it's all over if you did."

If those words were intended as a threat, they came only as relief. To think it could be over just like that. I looked around the café, where cross-legged professors sat reading and couples lingered over cups of coffee. To think that it could all be over.

"Did you go to Colorado?"

I noticed Dad's hard Midwestern pronunciation of Colorado.

"I don't think you had the nerve to take that trip," he said.

I couldn't tell if that was meant as disappointment or a challenge. So many emotions stirred in a single second, and in that second I understood more than I had in years. A rampage of freedom tore through my body: the wonderful freedom that's fueled by despair. I would walk away for good.

"Did you do it?" Dad asked.

My eyes were closed at one point. My eyes were wide open. My mind felt crushed, hollowed out.

"Did you so blatantly lie, Rachel?"

Blatant seemed an interesting choice of word when we'd been telling subtle lies all our lives. At least blatant lying could be distinguished, determined, and dealt with. Our lies were so deeply rooted we mistook them for truth.

"It's a simple question, Rachel."

It was not a simple question. There was nothing simple about knowing what needed to be done without knowing how. I'd never planned to turn my back on Dad's world all at once, but rather to inch myself away from him, gradually, naturally, over the course of many years and under the guise of normal lifecycle events: going off to college, or getting married and having children of my own. But it would never end if I didn't walk away, and although I hadn't worked out the details, like how I would support myself or where I'd go for holidays, the prospect of getting caught up in the lies made me feel like

I was on the top floor of a building that was slowly burning down.

"Dad," I said. "I went to Colorado."

The silence that followed sounded the way decay smells. My heart sped up. Nobody spoke. I repeated myself just to hear the way the truth sounded.

"Rachel," Mom said. "Please."

But there'd be no more favors. "I went to Colorado."

It was a silence full of dread, but a silence that finally put an end to the noise. I waited, because I knew what would happen next and I wanted to experience it happening. I knew it was the most important decision I'd ever made.

There was a click. I imagined Dad hung up first. And then there was another.

The outside air snuck under my sleeves and stung my skin. Car exhaust stained the snow, heaped up in the corners of the streets like mountains. Slush seeped inside my gym shoes. I walked up and down State Street in between the campus and the capitol. I walked until the sun began to set.

It was an unusually cold day in Wisconsin. I wanted to sit down at a bar and drink myself to a different place. In a way I felt like celebrating. I also felt oddly deserted, and that desertedness made room for everything that I wanted to someday feel. I knew, not in a fixed or stubborn way, but more in the way people say love is obvious when you meet the right person, that it was just right. I wasn't going back.

I returned to the thirty-six-person hippie co-op where I was living, cut through the frozen vegetable garden and down the wooden steps to the dock that hung over the lake. I heard nothing but the creaking of the deck, the thawing of thin ice underneath the trees.

It was that easy. That's what I kept telling myself, the details I could figure out, it was the vacancy I was not prepared to feel: a stillness infused with sadness that announced its permanence, tangled up inside my chest. There was a price to pay for being the one to leave.

SEVENTEEN

What happened after that was a succession of decisions that technically severed my connection to Dad. First he sent a certified letter declaring the termination of my health insurance at the end of the calendar year. Days later he reported the credit card he'd given me for schoolbooks stolen. I turned it over to the cashier. Last, he canceled payment on my tuition, which I learned my first day of class, when my name did not appear on the attendance rosters. I had spent the summer in Madison, training for a job as a residents' assistant in the dorms, which gave me free room and board and a stipend for tuition. I also took a night job as a waitress, a weekend job working with the elderly, and a tutoring job five days a week. Mom signed over $8,000 from a designated college fund Dad had started when I was young, and I invested it. The turning over of that money seemed to be our final settlement. I took the bus to Evanston and met Mom at the bank. We sat next to each other in silence as I signed the back of each check. "Now you've got what you need," she said. I said nothing, just walked her to her car feeling good and right about taking the money, thinking she was right, that now I had what I needed.

■

Toward the end of my senior year, as I started to think about getting myself somewhere else, I made my way to the university's mental health services. There I was assigned to several social workers before I met a woman I felt I could connect with. Tamar was in her forties, and, I assumed, a mother, from the photographs on her desk.

I walked into Tamar's office armed with clues to Dad's past. I was ready for some kind of deep exploration into what made Dad the way he was.

Tamar listened intently.

At the end of our second session, Tamar told me she didn't think she was qualified to talk about my father. I asked her what she meant. She said she didn't have the training. I told her I needed answers. She told me not to go there.

She sat back in her chair and looked up at the ceiling as if searching for an approach. Then she sat up straight, looked me in the eye, and told me I should wait until I was older, maybe even until I was married. "It just seems safer," she said. "To explore this when you're not available to him, when you have a life of your own, when you have the safety of a relationship that doesn't leave you exposed."

I nodded. The word *exposed* made me think of video games, of being blasted to pieces.

Tamar said to forget trying to figure out Dad. She said that all Dad had left of me was confined to his imagination. His imagination was his way of maintaining control. It was his way of keeping me close to him. I'd like to say my fear was then replaced by understanding, but that was not the way it worked. I wanted an explanation. Perhaps he was a sociopath. Perhaps he was bipolar. Perhaps if he'd gotten help twenty years ago, things would have been different.

I tried to let him go, to stop thinking about what it was he hated in me, but the effort merely made him seem to rise again. Not as my father but as a man with a mind and a heart that had beaten for twenty-five years before I'd come into the world. A man who was struggling *inside* more than anyone I'd ever known, a man whom I had gotten away from but hadn't let go of.

Tamar suggested we talk about Mom instead. We met once a week for the entire summer. We talked about what had happened in Mom's past that made her need Dad's approval so badly, that made her stay with him despite what was happening. I told Tamar about Grandma Janet and how she had married and divorced three times, spent most of Mom's childhood strung out on Valium, and completely neglected her responsibilities as a mother. Grandma Janet, as we had all witnessed over the years, was more interested in her hair and her fur coats than in what her children and grandchildren were doing in their lives. Mom's grandparents raised her. They drove her around, got involved with her school life, and supported her financially as she grew up, while Grandma Janet continued to make childish, selfish decisions.

I remembered a conversation we had years before, at a Chinese restaurant. Mom had told me Grandma Janet hadn't come to her wedding. I'd stopped eating and looked at Mom. Her cheeks were stuffed with egg roll.

"Grandma didn't come to your wedding?" I asked.

She hadn't come because she didn't like the way her own teeth looked. Mom got married in Chicago, surrounded by Dad's family. I asked Mom how that had affected her, or if it had at all.

"Grandma never grew up," she said. "Grandma's still a child. I wanted a father for my children."

As Mom often explained, that's why she stayed with Dad. Because she'd made a promise to herself, a commitment to get married and stick it out with the man that she married. She created the illusion of family, and pretended to make it work even when it wasn't working, so that she didn't have to admit to failing at the thing her mother failed at. Unlike her, Mom's children would have a father. Unlike her, Mom's children would have a family. Even if that father was cruel. Even if that family was a skeletal façade of what a family was supposed to be in Mom's imagination, a dream still caught in response to her own childhood. Mom didn't protect us because we weren't her priority. Although I disagreed with her decision, I could understand it, yet I couldn't help but see it as a question of character. And the older I got the less I could excuse it. Why hadn't Mom taken the time to grow up before she married Dad? Before she gave birth to me and Jenny? She was waiting to be saved, hoping to find a father in her husband, not just for her children but also for herself. It was true that Jenny and I saw Mom as another child stuck and silenced and wriggling to get out of trouble with Dad. But the fact was she could have left and she didn't.

In our last session together Tamar told me she'd never met anyone who'd used therapy so effectively. I left feeling high. And with that I found all kinds of excuses, ranging from lack of money to lack of time, for never going back to therapy. Perhaps my real fear at twenty-two was that a deeper understanding would take away my fuel. I was angry, less so with Dad than with Mom. Dad, whom I'd placed on a pedestal, whom I sought desperately, pathetically to understand and to love, was fading. It was Mom whom I needed to try to understand, not as my mother but as a woman, as a person who existed separately from me. It was something I couldn't accomplish until I got far away from her, geographically and emotionally, and from the person I became around her. I was full of judgment, couldn't stop thinking about all the things she didn't do that she should have done, and yet I still wanted her in my life.

EIGHTEEN

A month before I finished college, Mom called about a small favor. She asked if I would write Dad a formal letter of apology so she could come to my college graduation. I asked her what I was apologizing for.

"Better keep it general," she said, "general and typed."

I'd started setting boundaries that year. I was busy with school, working several waitress jobs and teaching so I'd have enough money to take off after graduation. I shared an apartment with my boyfriend, who picked me up from work late at night and cooked me dinner and brought me home to his family for the holidays. I knew by then what I would and wouldn't do for Mom.

"It's a setback, Mom, to say things I don't mean. I'm just learning how to say the things I do."

"It's for me, Rachel, so I can be at the graduation."

"Mom, I'm done with the craziness, I'm really done."

Mom was tending to things around the house. I could tell, because her voice hushed then echoed as she passed through certain rooms.

"This is not about Dad," she said. "You either want me there or you don't."

Mom phoned frequently, leaving messages on our machine. Messages like "Please don't deprive me of this, Rachel. It's very simple."

It wasn't simple. I was done apologizing. But I hadn't stopped playing the game. I sat on the phone with Mom, night after night, trying to get her to see it my way.

"You know how many bullshit apologies I've already written Dad?" I asked.

"So what's another one?" she asked.

It was that kind of thinking I wanted to stop making sense of.

At some point I caved and called Mom for the details. What exactly did Dad want in the letter? Again she said to keep it general. I told her I would think about it.

I could hear the relief in her voice. "Thank you for stepping up," she said. "Someone's got to be the adult."

She was talking about Dad and me. We were always talking about Dad and me. Mom had removed herself from the equation so long ago it didn't occur to her that she could have been the adult.

"He wants it to come from the heart," she said.

"But it's not from the heart."

We came at it from different angles. It was a technical issue for Mom, a simple letter I needed to write so she could be at my college graduation. For me it was a question of how much of myself I could give up for her.

"Well it should *sound* like it's coming from the heart."

I wondered what that sounded like to Mom.

"Dad is very sensitive right now."

I pictured him resting on a hospital bed, in a gown, after an accident or a surgery, not being able to talk, and Mom by his side, whispering, "Dad is very sensitive right now."

"He's not sensitive," I said, "he's sick."

Mom didn't like it when I preached. She was trying to come up with a reasonable solution for what she considered to be our problem.

"Should I start with my birth and work my way up?" I asked.

Mom laughed. I didn't. "See, this is why you *can* do this," she said, "because you've got a good sense of humor."

We had been laughing at what was not funny for years.

"Rachel, it's not just for me, it's for you. It's important to have your mother at your college graduation."

I did not write the letter. Instead I sent an invitation home. It had occurred to me that some part of Dad felt unwanted, and as much as I would have liked to absolve him of that, I couldn't do it at the price of lying, which was what the letter seemed to demand.

Mom called at all hours, often five or six times in a row. It was a week before graduation. Our answering machine was filled with messages begging me to write the letter. My boyfriend pulled the phone jack out of the wall each night before we went to bed.

We spoke the morning of graduation. She called at six a.m. to say that she was feeling sad. I felt less angry, more fragile, like the body underneath a water-resistant jacket that provided little dryness in a storm.

"But I have to make a choice," she said.

"You don't have to choose between your husband and your daughter."

"It's always been hard for me to make tough choices, Rachel. It's a difficult position he's put me in, a really difficult position, but I made a promise to myself when I married Dad, I promised myself that I'd stick with it."

Of course Mom had been making this choice all our lives,

but there was something in her verbal declaration, something that freed and frightened me. Mom was giving up on me, so I could give up on her.

"He's sick, Mom."

"You're right," she said, "he's a very sick man. But I made a promise to stick by him."

"We can change our minds, Mom. I mean, what's the good of having a mind if we can't change it?"

"You're young, Rachel."

"Don't tell me I'm young. You're confusing a promise with a choice."

"Well then, I've made a choice. It's not just a choice I'm making for Dad, it's a choice I'm making for myself."

My boyfriend was behind me, touching his lips to my hair. I hadn't heard him get out of bed. He was waiting for the coffee to percolate. I turned from the window and watched him going through his morning. This was real. He was real.

"You know, Mom. There are choices I'll have to make that I'm sure I can't even imagine right now, but you know what I am sure of? I'm sure that I will never marry a man who makes me choose between him and my child."

"I hope you never have to."

I felt a wash of relief. Mom knew what she was doing.

"Where's Dad?"

"New Orleans," she said, "medical conference."

"So get in the car!"

"I'm not going to sneak behind Dad's back."

"Why not? We've been doing it forever," I said.

I was afraid that she'd hang up, which at the moment seemed even more upsetting than her not being at the graduation. I wanted her to sit on that phone and wait for me to graduate. I wanted her to stay on while I walked the stage, took hold of my diploma.

I heard her sliding open the back door, letting out the dog. And it occurred to me that she was carrying on with her morning routine and that perhaps she hadn't been awake all night contemplating her decision not to come.

"Mom," I said, "you letting the dog out?"

"I am," she said softly, "the dog has a little diarrhea."

I lost it then, started pleading the way an eight-year-old pleads for ice cream.

"Come, Mom," I said.

"I'm making a hard decision."

"He's all the way in New Orleans."

"But it's a decision I have to make," she whispered. "Rach, I'm choosing Dad."

I can't remember who got off the phone with whom, but I remember the aching way in which I knew wanting her would ruin my life. Loving Mom was dangerous. It would hurt me in an almost physical way. She would not hurt me, but I would hurt myself in letting her love me, in loving her.

It wasn't just the graduation. I needed to get used to the idea that Mom might not be at my wedding or at the hospital when I had children or at the birthdays of those children or their high school graduations, their weddings. That was what I had to prepare for. That was what I had been preparing for my whole life.

My cousins came to graduation, along with my sister, Uncle Arthur and Aunt Jo Ann, and a handful of friends. We went to lunch afterward and when they left, I floated from one restaurant to the next, with friends whose families had invited me to join them. There was little time to feel sad. For a long time I'd been preparing for Mom's absence. And when I closed my eyes that night, it was relief that came over me. Thank God for this. It was exactly what I needed to clear Mom from duty. Like an honorable discharge, I could free Mom from what she wanted out of, and free myself from trying to keep her in. Later, I'd realize she wasn't dangerous at all. It was my unrealistic expectations of her that were dangerous. If Mom had been anything over the years it was consistent.

I was angry with her, but the anger felt like progress. Like the chills that occur before a fever breaks, it was a physical reaction to everything I'd repressed.

I worked four jobs that summer, with every intention of getting as far away from home as possible. Somewhere full of space and anonymity and opportunity, where I could get my head straight.

NINETEEN

California had a forgiving, gentle, quiet beauty. It was a separation of everything I knew from all that I wanted to become.

There were simple things I needed to learn. Things that seemed to be common sense for most of my friends. They'd been given some concept of entitlement from their parents that I was just then adopting for myself. They had ideas of how things should be: how men should love them, how they should be valued in a job or treated by a roommate.

I didn't know how to tell the truth. I'd become so accustomed to arranging my words around what I was supposed to say, or what I thought people wanted to hear, that basic communications were almost impossible for me. Saying "no" when I didn't want to do something, admitting to my own mistakes, asking for the things I wanted. The more settled I became in my living situations, in the traditions I developed with the friends who filled in as family, the more I let myself imagine a future.

Every holiday without Mom and Dad seemed like an assessment of my progress. I remember driving around Oakland in a borrowed car on Yom Kippur, with the Yellow Pages opened to *synagogues,* looking for a service I could sneak into. I had wrongly thought I'd be okay celebrating alone, on a blanket at Golden Gate Park, but I spent the day chasing after some form of community. I ended up sitting in a parking lot outside a huge Reform synagogue, watching families filter in and out. At sundown I drove over to a Thai place and broke my fast. For many years it was like that, the comforts that I wanted seemed lifetimes away, and still I felt immense relief that I was not at home.

My relationship with Mom and Dad was so much more clear-cut than most of my friends' relationships with their more functional parents, simply because I didn't have a safety net to fall back on. The men I loved had to be men I trusted far past any romantic inclination. I needed to admire them as people, the choices they were making, the way in which they thought and considered and acted upon everything. I needed my relationships with men to make me better and I needed to know I was making them better. It wasn't about finding a man who treated me well, but rather getting myself to a point where I could be good to myself, good with a man, good to a man. These relationships served as proof that I was on the right path.

While I was starting my life in San Francisco, Jenny was finishing college in Boston. Our infrequent talks rarely turned out well. Jenny would tell me that I'd turned into Dad, that I treated her the same way that he treated Mom. She knew that to me this was the single most hurtful thing she could say. And to her I said back, "You have the anger of Dad and the fear of Mom." It went on like this for some time, while we grew up and away from each other, because we had no context for one another outside of the maddening little universe we were trying to break out of. We were both struggling so hard to *not* become the two people that we hated and loved most in the world. But the most painful part of the estrangement was being out of touch with the one person who was hurting as much as I was, and in such a similar way.

Within the family, Jenny had been made invisible and I'd been made larger than life. Both of us went to work exaggerating these images of each other, and as we got older, neither one of us could recover from the image we'd created of the other. In such different ways, we'd absorbed parts of Mom and Dad that we could not get rid of. And when we looked at each other we could not see past those parts.

For years we stayed foreign to each other. Living on opposite coasts made it easy. We were relics, ruins of an empire that had died. Jenny wrote me letters to persuade me that my leaving home had been wrong. She called me one year, before Passover. "You've proven your strength," she said. "You've made it on your own. Why don't you just come home? You're hurting Mom."

I saw her calls as an insult to whoever it was I was trying to become. I didn't see that she missed me, or that her request for me to come home extended far past a holiday, far past Mom or Dad.

Jenny was moody, brilliant, and explosive and she was incredibly pissed off at me for leaving her behind.

When I was twenty-five, I got a letter from Mom bemoaning the fact that she hadn't come to see me run the Chicago Marathon the previous week and asking that I come home for Thanksgiving. She claimed to be in the process of divorcing Dad, again, and she promised he wouldn't be there.

Thanksgiving would be at Aunt Joan's, who was Mom's sister; just Mom and Aunt Joan and us girls. Would I please come spend the holiday with her and Jenny? There was no love lost between Aunt Joan and Dad, and I was relieved to hear that Aunt Joan had made it clear to Mom that Steve wasn't welcome.

I bought a ticket to Chicago. Nathalie picked me up from the airport the night before Thanksgiving. I slept at her house, ate breakfast with her parents, and locked the door behind them when they headed off to Michigan.

Morning turned to afternoon with no word from Mom, who had planned to pick me up on the way to Aunt Joan's. At two o'clock I picked up the phone and called Mom.

Jenny answered. "He's coming," she said.

"What do you mean he's coming?"

Every bone in my body felt crushed. I had the urge to hang up, but Jenny started in with the recognizable cry that erupted in her chest, and just hearing that sound brought back a distinct physical sensation of how it felt to have no escape.

"Put Mom on the phone," I said.

"You can't do anything about it," Jenny cried. "You know that."

Mom picked up. "Rachel, are you ready to go to Joan's?"

"Mom, you promised."

"Rachel. He has nowhere else to go."

I wanted to chuck the phone right through the big glass window in Nathalie's parents' family room. Blood rushed to my head and tears to my eyes, and it had nothing to do with Dad coming and everything to do with my own stupidity in returning home.

It hit me again that there was an enormous gap between what Mom wanted to do and what she was actually capable of doing. She was as comfortable lying as she was breathing. Her decisions were predetermined, made in the hopes of preserving a fantasy of family that had never existed.

She suggested I snap out of it if I wanted to have a nice

Thanksgiving, like a child should finish her dinner if she wanted dessert. Then she hung up. I got down on the floor with Nathalie's big German shepherd, buried my face in her fur, and wailed like an infant. After three years on my own, it was amazing how quickly I could be pulled back into that place. I lay there for a while, cursing myself for returning, getting back in the game when I thought I'd stopped playing, leaving the sanctuary I had in San Francisco.

Half an hour later Mom pulled into the driveway. I climbed in back with Jenny.

"You happy, Rachel? He's not coming."

Jenny took my hand.

"You got your way," Mom said.

"Could have been avoided if you'd kept your word," I said, knowing I was right but simultaneously feeling guilty for what Mom had coming. She would return home to an angry Dad who wouldn't talk to her for weeks.

We'd been at Joan's for half an hour when Dad walked through the door. Jenny and I sat on the basement floor playing a game of cards. I heard his voice, looked up the stairs, and saw his shoes. The same brown corduroy shoes he'd been wearing through the years.

Jenny smiled calmly. "I knew this would happen," she said.

I could tell she was scared for me, much more than she was for herself. I bit down on my lip; my mouth tasted of metal. There was no way to get out of there. I stood up, backed myself against the wall. Dad was not alone.

"Ali, this is Ellen," I heard him say, "and this is Ellen's mother, Janet."

"You shouldn't have done this, Steve," Mom said in the cross but sympathetic voice she used to scold the dog.

"Ali this is Ellen's sister, Joan."

"What are you doing here, Steve?" Joan asked. "I specifically told you this morning you weren't invited."

"Oh," Dad said, sounding hurt. Then more sheepishly, "I thought we'd just stop by."

Being in such close proximity was paralyzing. I walked into the dank, unlit back room of the basement, with the tool kits and the laundry machine, the cobwebs and extension cords. Dad was making his way down the stairs, slow and steady creaking. I placed myself behind the water heater, which was tall enough to block my body; half wishing I'd stayed where I was instead of hiding like a coward. The sound of his voice hit the room like cold November air smacking my face.

"Rachel? Where are you?"

I sucked in my breath. He stood about ten feet away from the water heater.

"I see you, Rachel."

His camera was dangling from his neck. "You look good, Rachel, really good."

I hoped that he was kidding. My eyes were ripe with tears, my nose wet with snot.

Mom thought I had a choice: I could get on with it and have a nice Thanksgiving or I could be angry. But she didn't realize that standing in the corner of Aunt Joan's basement, hiding from Dad, at twenty-five, was far from one of the choices I wanted to be making.

"Should I take a picture?" Dad asked, holding his camera up to his face. "Stay still."

"Jenny," he said, motioning for her, "Jenny, join Rachel for a picture."

She stood next to me, both of us smiling, both of us crying. Then she put her arm around my shoulders.

Being chosen in our family held a lot of unwanted weight. Dad had chosen me. Mom chose Dad, and Jenny had stuck it out on her own.

Dad shot a couple of us together, and a couple of me alone, and then he slung his camera back around his neck and moved in closer. Putting his arms around me quickly, he immediately retracted, stepping aside with his head hanging low and his arms dangling like he didn't know what to do with himself. Then he backed away, stood rocking onto his heels with his arms across his chest.

I'd given up on Mom being a mother according to my definition of what a mother should be. I no longer expected things of her, but I did have expectations of Jenny. Jenny and I had a relationship with one another, a complex relationship but a relationship nevertheless. And Jenny was involved in a way Mom never was. She *wanted* to be my sister.

Jenny's eyes were wide open. I saw in her face the rest of our lives. She looked at me with caution, and I knew we were feeling something similar. We were terrified of losing each other.

"I've brought a friend from the hospital," Dad said. "He'd really like to meet you."

Jenny stepped away from the water heater and held her hand out to me.

"This is so crazy," she said, laughing and shaking her head.

I laughed because it was a craziness we shared, one I'm not sure I could have gotten through on my own. I took Jenny's hand and together we followed Dad upstairs to the kitchen.

Years later I'd come across those awkward pictures Dad took and Jenny and I would laugh that unstoppable, gut-wrenching laughter that is and always will be the most real thing we have together, because it is ours and because our family is funny to us in a very particular and bizarre way. Because we both understand that humor is a distorted savior and a deeply sad cry.

Ali, Dad's guest, was a slender, chocolate-brown man with a mousy face and a shiny black mustache, who'd come to the U.S. from Pakistan. He stood in the kitchen, unassuming, in a white-and-blue pinstriped button-down shirt.

Dad had a habit of adopting small, dark, non-English-speaking men who worked at the hospital. Dad was very nice to them, gave them rides and invited them to our house for the holidays. He was interested in their lives; inquired, empathized, and shared his own stories of being in their countries.

I looked at Mom and then at Jenny. Their aching faces seemed to say "stop" and "go" all at once. They didn't know what to do, and for a second I thought they were waiting for me to speak up. I wished that Ali hadn't been there, a psychological shield for Dad who assumed I'd revert to faking it because he'd brought a guest. It wasn't fair that Dad had shown up. In war it's prohibited for combatants to pose as civilians.

"Ali," I said, loud and slow, not because of the language barrier but because my own voice sounded like it needed urging, "I'm sorry if this is uncomfortable for you, but Steve is a very sick man."

Dad wore a post-Novocain smile. Mom's face went blank. She walked toward the stove with a potholder in one hand and a mixing spoon in the other. Ali stared at the white linoleum. Everyone seemed to be patiently waiting it out, like a sudden and short rainstorm in the middle of a perfectly clear day.

"He was asked not to come here today."

Aunt Joan began to nod in agreement.

"I haven't seen my dad in three years."

This was what got a reaction out of Ali, whose lips went from a smile to an O shape. He looked at Dad for confirmation.

"See, we're used to letting Dad control things. We're used to just shutting up and acting how we're supposed to act for Dad, for Dad's guests."

Mom gave me a look that begged me to quit.

"He's not a good dad, Ali. A good doctor, definitely . . . but not a good dad."

I turned to look at Dad. I wanted him to know that I was playing fair. But he wasn't looking at me. He was looking at the floor.

"Ali," he said, "Rachel doesn't seem to like me very much."

We all sat down to eat. Dishes of sweet potatoes, string beans, and cranberries were passed around the table. At one point, Dad stood up to take a picture. "My house," said Aunt Joan, "I'll take the pictures."

"There's vegetarian stuffing for Stevie and Rachel," Mom said. And there was not a hint of regret in her voice. We were together as a family, for the first time in years, which, despite the skewed nature of our togetherness, was exactly what Mom had wanted.

By the end of the meal, an exhausted quiet had fallen over the table. Dad and Ali were the first to leave. I made a point of shaking Ali's hand. It was a stupid point to make, but it was my way of apologizing for causing him discomfort. I saw Dad fiddling around with the zipper on his jacket, and I thought to tell him what he'd told me many times before: "Don't force it."

An awful miscommunication—that's what Mom called our Thanksgiving.

"He's not well, can't you see?" I insisted. "I don't want to be near him." I needed Mom to understand that this wasn't about me nursing a grudge.

"But he's your father," Mom said. "And he's always going to be your father."

Mom reached for my hand.

"You shouldn't have let him come," I said.

"I had no choice," she said.

I thought of Thanksgiving the year before. Pots and pans spread out on the floor of our barely furnished apartment, where I'd stayed up with my roommates playing poker, drinking whiskey, eating bastardized versions of our family's recipes. For one reason or another we were not with our families, and there existed an understanding between us; a random scattering of Midwestern transplants making our lives in San Francisco.

I had a choice.

It would be many years before Jenny and I were far enough away to look back at our past together, but I returned to San Francisco determined to make our relationship work. As far as I wanted to get from home, I couldn't afford to isolate myself from Jenny.

TWENTY

On the airplane back to San Francisco, I rewrote the rules I'd need to protect myself and get on with life. I could, under no circumstance, trust Mom. I could not participate in her world if I expected to create my own.

Dad could kill a man and Mom would stand by him. It had nothing to do with Dad and everything to do with the image Mom chose to see. And by then I knew that if I expected anything at all of Mom, I too would be choosing to live dishonestly, in a similar self-perpetuated fantasy land.

In the years that followed I cut myself off from her completely. Not by pretending she didn't exist but in truly abandoning any expectations I had of her. Included in that was abandoning any expectation I had of myself to be there for her. I forced myself to see Mom as an adult. I didn't owe her anything because I'd escaped and she hadn't. That was her choice, one that would ultimately limit the depths of our relationship. I regularly ran myself through scenarios, practice drills, of what I would do if Mom called with new plans to leave Dad. What if Dad got in a car accident? What if he left Mom? What if she moved to California to start a new life? The answer remained the same. I could not afford to give up any part of my life to take care of Mom. At least not then, when I was just growing into myself.

She sent letters throughout the years, pleas to “come home.” She sent photographs of Dad holding me on my first birthday, me at age two riding on Dad’s shoulders, me at age nine standing shyly behind Dad in the snow. She sent silly books that trivialized our situation. Books like *Don’t Sweat the Small Stuff with Your Family*.

And I accepted them. Reluctantly at first, but with the understanding that if we were going to be in each other’s lives, I had to learn to receive what she could give. I wanted to accept her and I didn’t know how. I knew it would take years to recognize in Mom exactly what Dad could not recognize in me. She saw things the way she chose to and not the way I’d wanted her to. And her ways of mothering contradicted my very firm ideas of what a mother, by nature, should do for her children. But over the course of the next five years I’d stop holding her to my own standards of right and wrong.

Once I could see Mom for who she was, and not the mother she had failed to be, I’d be able to forgive her. She was her own person, not just my mother, and it was this person I was forging a relationship with.

What I couldn’t forgive was the façade Mom kept up after all that had happened. Mom was willing to compromise everything so she wouldn’t look bad. As I’ve gotten older, my desire to reconnect with family friends has intensified. Mom gave me short updates when I asked about the kids I’d grown up with but hadn’t seen in ten years, but what I never asked, and always wanted to know, was where they thought I’d gone.

A decade later, when I eventually reconnected with these

children of our family friends, I saw how intricately Mom and Dad created their own reality. Only now am I beginning to understand what others must have seen and heard and settled on as "the truth."

"It was just known," said one of these friends. "We were not allowed to mention you. For the last ten years your name has not come up at the table."

I was devastated to hear this, and a little surprised. I thought for sure Mom would have let them know what I was up to, if only to paint the picture that she and I were in touch.

"I asked about you a couple of times and my mom would kick me under the table. It was kind of like we were all supposed to accept the fact that you had disappeared."

I was curious to know what she made of it. Did she think I was in trouble, homeless, on drugs?

"I don't know," said the friend. "No one had any idea what to think. My parents said you gave your parents a hard time growing up, that you had problems, and I guess that was kind of what we went with when you disappeared."

Hearing this validated my mistrust. When I stopped expecting things of Mom, I mistakenly thought that the one thing she would give me in return was permission to tell the truth.

At last I understood why no one intervened. I had had no idea how thoroughly Mom and Dad covered their tracks. Just recently my cousins asked about Mom's political connection in D.C. They were shocked to find out that the Page Program was something I'd applied for. They'd heard that Mom and Dad needed a break from me, that I had gotten out of control at home, and that Mom had a friend in D.C. who'd helped me into the Page Program.

Slowly, judiciously, I began to reach out in small ways to

let old friends know I was okay, and in doing so I hoped that someone would show some curiosity in what had gone on and what was still going on in our family. One year I ran a marathon in Anchorage, Alaska, for an organization that raised money for leukemia research. I sent letters to everyone I knew to raise the money I needed, including Dad's entire family. The cousins and aunts and uncles in Minneapolis, who saw Dad as a saint, the family I hadn't seen in years. I remember how self-consciously I stamped those envelopes, wondering exactly what they would think when they got that letter asking for money. Jenny immediately berated me for soliciting money from Dad's side of the family after I'd abandoned Dad. Plus, I imagined them thinking I was spending the money on drugs or rent rather than leukemia. And so I started including information on the organization and tax-deductible forms. I hoped that they would donate, but so much more than that, I hoped that they would know I'd done well for myself, and question why I'd left.

■

When I was twenty-five, I joined a program to become a court-appointed special advocate for a girl in foster care. As part of my training to become certified in California, we had weekly speakers lecturing on the foster care system. Jenny had recently graduated from college in Boston and moved back to Chicago. Despite having acknowledged that we needed each other, we also needed to move on, and every conversation, every visit, brought back a slew of memories we were trying to forget.

One night the seminar was given by a specialist who was pointing out the difference between children who'd been abused and children who'd been neglected. Neglected children felt invisible, as if their presence had no bearing on anyone or anything. Abused children felt all too visible, as if they were the center of everyone's world, because they had been the center of someone's world, the recipients of an abnormal amount of attention.

I thought about my own ridiculous attachment to what others might be thinking of me. When I chose a seat on a bus, I often thought the people I had not sat next to were taking it personally, that they'd assume I'd avoided them because of their race or their weight or their age. Thoughts like this churned and reeled in my brain until I talked myself out of what I knew was a peculiar form of narcissism.

And it wasn't just with strangers; it extended to jobs and relationships. The idea that I could not quit an assistant job because everyone was depending on me, that my boyfriends would be devastated without me in their lives. I had an incredibly inflated concept of my effect on others, a misconception that trapped no one but me.

One night, a friend dropped me off at the apartment I shared with six random people on Ashbury Street. I had forgotten my keys. Because it was after ten o'clock, I refused to ring the bell. I climbed back into my friend's car and told her I had to spend the night at her house.

"Well, didn't you ring the bell?" she asked.

I told her I wasn't going to ring the bell. It was after ten and I was in the wrong. She assured me that my roommates were probably awake and, even if they weren't, one of them would come to the door and go back to sleep. It wasn't that big of a deal. Instead of ringing the bell I did several laps around the house, trying to determine if anyone was awake or if there was an open window I could climb into. My friend was getting annoyed. When I got back in the car I said I'd rung the bell and no one answered and could I please just sleep on her couch.

"How many times did you ring the bell?" she asked.

"Trust me. Enough."

She didn't trust me. She insisted on getting out of the car and ringing it herself, until I broke down in tears and begged her not to.

"Why'd you lie?" she asked.

The tears were really coming down. I just wanted to curl up on her couch and go to sleep. It was a Sunday and both of us had to work the next day and I was dressed in shorts and hiking boots, which meant I would have to borrow work clothes from her. We sat in the car for a while. She was annoyed and I had shut down. All I knew was she didn't get it. She didn't get that I was in the wrong and I had to pay the price. I didn't get just how much of Dad's thinking I'd accepted as my own.

"You can sleep on my couch. You can sleep on my bed. It's not that. It's just so stupid that you can never be wrong, Rachel."

"But I am wrong," I said. "And now I can't go home."

"Shit happens," she said. "You're not gonna ruin anyone's life."

I hadn't seen it like that. I had come to know my fear as martyrdom. I went back to Erika's, crawled into her sleeping bag on the couch, where I lay awake until one in the morning. Then I put my hiking boots back on, walked to my apartment, and sat on the porch for a good half an hour before I rang the bell.

"No worries," was what my roommate said when he came to the door in his bathrobe.

For days I avoided going home when my roommates were

around. I'd convinced myself they thought I was negligent and selfish. I'd opened the door late at night for them, but it seemed an entirely different situation when I was standing outside without my key. I, who was larger than life. I, who would ruin their night, leave each of them awake for hours, lying on their backs thinking of me, hating me, knowing how irresponsible I was for forgetting the key, how selfish I was for waking them up. Overcoming the internalized belief that I was at the center of everybody's world, that I could ruin somebody's life, was my biggest battle.

And yet, talking about boyfriends and classes and jobs, Jenny constantly said, "They don't hear me," and "They don't see me," and "No one even notices I exist."

She yelled and cried and threw fits. It was something I envied, her capacity to make noise, her desire to be seen, her determination to communicate how she felt. I had remained and continued to remain silent. When I was hurt, when I felt wronged, when my friendships and relationships depended on my speaking, I shut up and retreated inside myself. Now I was trapped by what had once saved me.

After the training session on abuse and neglect, I called Jenny. That phone conversation was the beginning of something. It was the first time we'd talked about how it was for us to watch the other. What I didn't know was that Jenny swam in an enormous pool of guilt because of the way Dad treated me. That she worried about me in an almost maternal way.

It wasn't easy; it's still not. She needed my constant assurance that I wouldn't leave her, as I needed hers. But what Jenny and I had was a shot at a real relationship. We spoke on the phone, often working through things both of us were still angry about. Jenny explained what it was like to sit through years of conversations that didn't involve or acknowledge her existence, how badly she wanted to be my friend, and how removed I was from her and from everyone. A lot of it was normal stuff, older sister/younger sister resentment, but what both of us were afraid to say was that by leaving home I forced Jenny to become an only child. She felt she had to make up for my absence. When conversations touched on this raw nerve, we'd retreat into our old habits. I would grow silent and Jenny would throw fits. She'd scream and cry and say things that hurt so much I'd simply put the phone down and begin to fold my laundry. "Are you done?" I'd ask after a while.

"Don't hang up on me," she'd finally scream. "That's how you live. You walk away from everything."

I stopped hanging up. After a while, Jenny came to understand that I wasn't going to leave her again. She flew to San Francisco for visits. They weren't always good, but they were solid attempts at understanding each other.

One night Jenny called me after she'd returned home from a friend's wedding. I'd been through a wave of friends' weddings and Jenny wanted to know if I got sad seeing a parade of brides with their fathers and their families. I told her that I used to, the first couple of times, but it got better. It got better as I got older and relinquished the hopes of having or finding in someone what I hadn't found in Mom and Dad.

"At some point something shifts and you stop wanting what's missing and start wanting the things you create for yourself," I said.

"But when you see those dads walking their daughters down the aisle, don't you feel like you could die?"

"That'll go away," I said.

"Really?"

"I promise."

I could hear in her voice that she believed me.

"Desires change, you know. I used to hope I'd wind up marrying someone with perfect parents: the mother-in-law that takes you in as her own daughter, the father-in-law who adores you. But that's not up to us."

"It would be nice," Jenny said.

"So nice," I agreed. "But at some point we'll have to become what we want. We'll have to be *that* mom for someone else, hope we marry a man who becomes *that* dad. We can't just go on wishing for some concept of family that doesn't exist."

It was not a conversation about other people's weddings and it was not a conversation about what we hadn't had, but a promise to ourselves that we'd stick by the other.

Jenny's experience was different from mine. Her version of the story is her own, but of all the people in my life her version is the closest to mine. We need each other for reasons beyond our shared past, but there's a burdensome truth, a complication that requires each of us to do whatever it takes to relate to one another. Only we know what really happened, that what happened was not made up or exaggerated. Jenny knows me better than anyone. She had always been there, quietly watching and witnessing.

Six years after I stopped going home, Jenny followed suit. It happened over something minuscule, like my trip to Colorado, and it happened over the phone, as well. When Jenny called to tell me about it, she was in a rage. "I'm never talking to him again!" she yelled.

I didn't care if she did or she didn't, all I wanted Jenny to see was that it had nothing to do with her. The world Dad lived in, and subsequently the world Mom chose to hide in, had nothing to do with any decisions Jenny made. She could have whatever relationship she wanted with either one of them. Just like I could. The relationship was entirely in Jenny's control. Because there was nothing unconditional in our parents' relationships to us, the conditions were entirely ours to determine.

I told Jenny that the actual incident would blow over. It seemed like a big deal, but it was only a big deal to her. "Mom, Dad, they won't remember it. I promise."

And that was exactly the problem.

Jenny was supposed to pick them up at the airport on their way back from Cancún. A week before they returned Jenny remembered she had an engagement party that same night. She called Mom, who said they'd get a cab. But after talking to Dad, Mom called back to say Jenny had made a commitment and she'd broken her word. She'd broken their trust. It went on like this for a while, both Jenny and Mom all worked up, calling each other, fighting it out, hanging up angry. But it was what Mom repeated over and over again that made the two of us erupt in laughter.

"If you're not at the airport Jenny, it's all over. Really, it's all over."

And so, in the same nonchalant manner, Jenny stepped out of Dad's life. And after a year had passed, Mom, with a roll of her eyes, would say, "Jenny's declared war on Dad, too."

Jenny could return in a month or ten years or never; she will when she wants to. That's what I told Mom. And then, as if to assure her that she would get back one of her children, I said, "Don't worry. She'll be home soon."

At times I'm surprised to see how much Jenny, like Mom, has bought into the façade of a perfect family, despite everything we've been through.

When Jenny got engaged, she told me her fiancé, Dave, had asked for Dad's permission. I tried to keep my voice even when I said, "Dave's pretty traditional, huh?"

Jenny laughed. Then she said something that reminded me how differently we saw our lives and ourselves.

"Rachel, any guy who has any respect for you is going to ask for Dad's permission before he asks you to marry him, regardless of the situation."

For me there was no situation and there was no Dad. For Jenny there were both.

Jenny still hoped for and expected an illusion similar to Mom's. But what I knew, I knew only for myself. I couldn't make Jenny come to the same realizations I had. We were made of different materials and dealt with our experiences accordingly.

I was not surprised when Dad refused to go to Jenny's wedding.

"It's not you Dad has a problem with," Mom said, as if this information was supposed to reassure me. "It's Arthur and Debbie and Jill."

Dad and Arthur had had a falling out several years before.

"Dad would *actually* like to see you," Mom continued.

"Trust me, I know," I said.

Nor was I surprised when Jenny took great satisfaction in Dave's calling, e-mailing, and faxing Dad. "He's horrified that a father would not attend his own daughter's wedding," she said.

How much had she told him? How much did he believe?

But Jenny was looking for the same validation that I'd needed. The reassurance from someone she loved, who was outside of our reality, that something was terribly wrong. Wrong enough to justify her decision not to go back. She was waiting for the world to offer her a better alternative.

TWENTY-ONE

My friend Nicki slowed the car as we neared my parents' house. We parked on thc opposite side of the street. She shut off her lights and turned on the heat, and through the driver's-side window we sat and watched the conclusion of Dad's Passover Seder.

"This makes me kind of sad," Nicki said. "These annual drive-bys."

Ten years had passed since I'd been home for a holiday, and still I insisted on riding by and watching what I'd left.

"What if I rang the bell?" I asked.

"They'd probably think you were Elijah." Nicki stared straight ahead.

I recalled the childish lethargy I'd felt as a kid toward the end of those Seders, back when the living room was filled with aunts and uncles and cousins and close family friends. Back when Dad used to sling me over his shoulder and carry me up to my room. Slowly, over the years, people stopped coming. The family friends complained that Dad's five-hour service was too long. Our immediate family followed.

"Dad's excommunicated everyone from the Passover Seder,"

I said to Mom the year I left for college. "It's like survival of the fittest."

Mom shook her head. "I know. It was so much fun for a while."

The living room glowed golden orange, like a late afternoon sun. I could feel the warmth they'd created for their guests, and how that warmth would carry those guests through the evening as they got into their cars and drove themselves home. I tried to imagine the admiration Dad's students were feeling for him, what they would say to their spouses as they drove away. I've always liked to think I had a certain insight into other people's character. Would I have had that with Dad? Sitting as a guest at his table? Would I have wondered where his daughters had gone? I assumed they knew he had daughters. There were pictures all over the house.

The living room was full of Dad's medical students.

Nineteen guests. That's what Mom had told me. "It's going to be a good one," she'd said. "A really nice crowd."

Dad's chosen ones.

"Some Jews, some Catholics, some Hindus and even a Jain," Mom had said. I imagined they were flattered to have been invited to their professor's for a holiday. I would have been.

I remembered finding a Christmas card in our basement from one of Dad's med school students.

The card read, "Dear Dr. Sontag, Thank you for being so much more than a teacher."

■

The versions of ourselves we present to the world are perhaps the versions of ourselves we most want others to know. We split and divide at the core, recreating ourselves, until we determine the perception we best like. To the student who wrote the card, Dad must have been something of a hero. Finding that card gave me an odd sense of relief that Dad had found in others the adoration that I could never give him.

Nicki rested her elbows on the steering wheel. It was cold and I had my hands tucked into the sleeves of my jacket.

"Just a few more minutes," I said.

In the solitary manner of watching from a certain measured distance, far enough to be in the audience but close enough to feel some sort of engagement with the actors, I watched as my parents and their guests inhabited the traditions of family. Figures walked stage right, stage left. Mom's head of curly hair; Dad's slow, concentrated crossing of the living room, huge smile, eyes averted toward the ground. It was like watching a play I'd seen before, having a general sense of plot and characters, which allowed me to see more deeply the second time around.

I figured Mom would clear the table, soak the pans in the sink, and load the dishes in the dishwasher. Dad would collect the supplemental readings he'd put together and store them in a box for next year's Seder. He would leave the tablecloth on the table. She'd deal with it tomorrow. We never cleaned up all the way.

"Maybe we should drive the car right through the house," I said.

Nicki didn't laugh.

I'd spent many years trying to get away from Dad. And then I did.

That was my story and that was where I wanted it to end. But it wasn't over. I still said things like, "Maybe we should drive the car right through the house," things that made me sound more angry than sad, but sad was what I felt. I was sad knowing that I could never go back. That one who leaves must eventually be the one to return. That returning would not be reconciling with Mom and Dad but arriving in a place where I was creating ways to live, not just trying to survive.

■

It still makes me uneasy to explain our estrangement. Sitting next to someone on an airplane, or across the table at a dinner party, or beside a man I'm falling for, I will explain, when it comes up, that it's a difficult relationship, that my father is a complicated man.

An onslaught of inquiry follows. Are both my parents Jewish? Where in Chicago do they live? Are they still together?

Yes, I say. They were both raised Jewish. We grew up on the south side of Evanston. That's where Northwestern University is, a beautiful place to raise children. Yes, they're still together. I know. It's rare. They're happy.

Younger people react differently. Older people are often parents or grandparents, they have lived longer and they know what matters in life. They say that it's a shame for me. I'm not punishing my father. Life is short and this is my loss.

"What if he dies?" they ask.

It is a question I'm not supposed to answer but rather to consider. But what I'm being asked to consider is another person's regret, not my own circumstance. It is easy to tell when the person asking the question has lost a parent with whom he or she hadn't made peace.

"Have you thought about that? What if he dies?"

Sometimes I look off in the distance as if to demonstrate that

I am considering this for the first time, that my father might die, and that the asker of the question has raised a good point, one that I haven't fully examined. Depending on the person, I will answer. Yes, I have considered this possibility.

I am told to make peace with him now because there's no bringing him back once he's gone. But, of course, I had never been at peace with Dad. The peace I'd found was in stopping our struggle, coming to terms with the parts of Dad that lived inside me. Forgiving Mom, accepting her love for Dad and the love she had left for her children. Accepting that in an imperfect way, each of us has found what we need.

I thought about the day my college therapist, Tamar, told me she wasn't qualified to talk about my father, the disappointment I felt in having to dig into my relationship with Mom. Mom had hurt me in ways Dad wasn't even capable of. Mom was real, human, and conscious of her decisions. Mom has become more real to me over the years, whereas Dad's impact has faded. My relationship with Mom is deeper, more complicated, because we have a future together.

Yet our relationship demands that I monitor my emotions and expectations. I don't want to give, want, or need too much. I can't let my guard down by trusting her. We simply preserve what we have. And while it's not ideal, it is alive.

There will never be that sense of the unconditional. Mom is not the person I turn to when I need something. Hers is not the name I write down on emergency contact forms.

Loving Mom means not only accepting her love for Dad, but also forgiving him enough to help Mom brainstorm ideas for his sixtieth birthday or their upcoming anniversary. Loving Mom means going along with her fantastical creation of reality, something I'll do to the extent that it doesn't involve me.

She calls me with thrilling accounts of their trips to Cancún. The snorkeling, the boat rides. They travel. They go to concerts in the summer. They watch movies and meet their friends for dinner. Mom has the marriage she's always wanted.

On one of Mom's visits to New York, where I moved when I was twenty-eight, we walked into a candy store in Greenwich Village. Mom wanted chocolate and I wanted licorice.

"Licorice? You want licorice? Real black licorice?"

She gave a good laugh and looked at the guy behind the counter.

"I can't believe it. She and her father, the only two people in the world who like real black licorice!"

"Licorice and Butterfinger bars," she said as we left the store. "You are your father's daughter."

"Mom," I said, "c'mon."

"And your teeth. I mean whose genes do you think you got? All those cavities and root canals."

"All right," I said. "Enough. I could have inherited worse things from Dad. Things I couldn't have filled or extracted."

"That's true," Mom said, nodding intently.

There is always a conversation being had over and under the ones we actually have.

■

Back in the car with Nicki, I realized the holiday drive-by had become a humiliating habit. I'd come to enjoy the relief of being able to leave despite the disappointment of having to. Mom and Dad were going through the motions of hosting a Passover Seder, and I was going through the motions of revisiting what I'd left long ago, trying to retain a physical memory of the place I could never return to.

Nicki's windows were fogged by breath and heat. It was one in the morning, the last time I'd need to make that trip. The next day I'd go back to New York. Back to the only apartment I'd ever lived in longer than two years. But right then I was home, the only home I had, the home I'd always have: the place inside of me where I wasn't struggling to prove my strength to strangers, to lovers, to myself.

Nicki started the car. I could feel the motion of driving before we pulled away, the cold seeping in from outside, the swell of relief that came over me knowing I wouldn't be back.

ACKNOWLEDGMENTS

I am very grateful to Amanda Urban, Lee Boudreaux, Dan Halpern, and everyone at Ecco Press for your belief in this book. I feel honored to have worked with you. Thank you to Derek Loosvelt, Moriah Cleveland, and Keenan Schofield for your friendship, edits, and insights. To Dan and Erika Henschel for Colorado when I needed it. Over the last six years, my mother has graciously allowed me to read her almost the entire book over the phone. Thank you for laughing when it was funny and listening when it wasn't. You supported me in both these reactions. Love and thanks to my entire family. I know this subject matter is personal, I just didn't know how to live without sharing it.

About the author

About the book

Read on

Insights,
Interviews
& More . . .

About the author

Meet Rachel Sontag

Antonia Wright

Rachel Sontag was born and raised in Evanston, Illinois. She received her MFA in creative writing from the New School. She lives in New York City. This is her first book.

More Than Just Myself

I AVERAGED several solid two-year relationships over the course of my twenties: each of them with good men who I didn't want to marry. I was young and attracted to change and freedom, and to the transience of being in a perpetually uncertain state. I was uncertain about what I wanted out of a job, in a man, of myself. I was uncertain of the city I wanted to live in, where I'd find financial stability, or even if I wanted that financial stability. There was a comfort I took in discomfort, in surviving, in making it on my own.

I was nine the first time I left the country. It was a family trip to Morocco, where we rented a small, un-air-conditioned Ford Fiesta and drove across the country. It must have been the combination of the Saharan August heat and the dirty water I'd accidentally drunk that caused me to vomit for the majority of our trip. While I have a few soft memories of the marketplace in Marakesh, the pink-painted cities and basins of colorful dyes in Fez, the thing that I remember most was just how close Dad stood behind me while I vomited. There were no toilets, just holes in the ground. Afterwards I filled a pail with water and washed the vomit down the hole. "You think you're sick?" Dad asked. "This is nothing. You should see how people live in the villages of India." Then he took his hand and placed it on the small of my neck and I could feel how proud he was of me.

Most of our trips started out the same way, with Dad in the front of the cab on the way to the airport asking, "Do you kids know anyone else that travels the way we do?"

We stayed in youth hostels, slept in rental cars, and ate dinner in the parking lots of grocery stores. We memorized state capitals, learned how to trade money on the black market, got strip-searched and detained at border crossings. I came to romanticize ▶

More Than Just Myself *(continued)*

discomfort, considered it essential in experiencing the kind of life-changing adventure I was seeking.

And I'd been blessed with an independent spirit, so the natural thing to do at the end of any break-up was to quit my job and leave the country, or at the very least, the state of California. After one relationship I packed up my tent, bought some maps, and went to spend the summer in Alaska. After Mike and I broke up, I cashed in my frequent-flyer miles and headed to Central America. I was twenty-seven, floating from one job to the next, trying to support a writing habit.

Mike managed the bar of a restaurant, a position that kept him drunk and out at all hours, a position he insisted was temporary, like his drinking. He wanted out of what was temporary: nights that made him go fierce and then blank, nights in which he lost his money and his wallet and his mind. Nights I'd piece together for him the next morning with the remains of clues he'd left.

Mike was not always drunk. When he wasn't we were biking in Marin, making our way through various art galleries, lying on his rug watching old Woody Allen movies. Mike cooked monkfish and mashed potatoes with rosemary, frittatas and pancakes and soups. He'd race me down the beach, push my face into the water fountain when I bent down for a sip, licked the sweat off my arm. He liked that I woke up early in the morning, that I made him quit smoking and vote in the local election.

Like a person becoming proficient in a second language in order to better understand a different culture while living abroad, I concentrated hard on ways to successfully date a drunk without losing hold of myself. The more I drank the more sense Mike seemed to make, reaching for the aspirin in the morning and each other late at night.

He didn't want me to tolerate his drinking. "Come down hard on me," he said. "I don't want to drink like this, I don't."

But I didn't necessarily need him to stop. I enjoyed the dramatics of dating a drunk. I saw our relationship the same way that he saw his drinking: as a temporary deviation, something to slow down and seal up the time between who I was and whoever I was bound to become.

When I arrived at San Jose International Airport I exchanged dollars for colones and bought a pack of tampons at the duty-free shop before boarding a bus down south to Quepos. Five hours later, after being dropped off in the city's center close to a strip of cheap hotels, I rented a room at the Mar Y Luna Hostel equipped with a small bed and a ceiling fan for four dollars a night.

I found an empty bar across the street, downed several bottles of Imperial Beer, and then stumbled back to the hostel. Abruptly and in the middle of the night, I woke in an almost feverish sweat. There were cracks in the ceiling, yellowish plaster peeling off the walls. In an attempt to cool my body, I soaked towels in the sink down the hall, placed them on my mattress and got back into bed.

I told myself to relax. My body was reacting, rebelling, acclimating to cultural change, climate change. That was a well-known fact. Another well-known fact: When birth control is used correctly one should not get pregnant. I pressed my hands against my belly.

I'd been taking that pill for four years. Same time. Every morning. How many of those mornings I'd looked down at the pill in the palm of my hand, considered the size, wondered how something so miniscule could prevent something so massive.

I watched a large crab run the length of my cement floor. I rolled onto my stomach. ▶

More Than Just Myself ***(continued)***

Days passed, each morning I woke before sunrise and stood staring into a bowl of clear toilet water. I replayed the scene of Mike dropping me off at the airport. An odd, quiet togetherness solidified between us. He had double-parked his car, removed my backpack from the trunk and placed it on my back. Then he'd handed me a copy of Richard Adams's *Watership Down*. I'd looked at the bunny rabbits hopping around on the cover.

"Don't dismiss the book. It's not about rabbits."

I'd buckled my backpack around my waist.

"And don't dismiss me," he said.

I couldn't dismiss Mike. Part of him was growing inside of me. But I also didn't have it in me to pick up the phone and call him. A week passed and I began to feel pregnant, and because I didn't immediately want to deal with that pregnancy I distracted myself with the technicalities of journeying. I headed up north to Liberia, a medium-size town an hour south of Nicaragua, which took me a day and several buses to reach. My mind felt disassembled from my body. I blamed it on the heat. My face bled with oil, with grease that smelled more like Wesson than sweat. I removed the soaking-wet towels I'd slept on, uncertain of where I was until I placed my feet on the cool concrete floor. And then it came to me. I was nauseated, every morning, acutely aware that this was really happening.

I have always thought that if I adhered to certain rules, I'd be rewarded. But I brushed my teeth and still got cavities and I took those pills and still got pregnant. Day after day, I waited for this pregnancy to fix itself, for God to say, "Sorry, you took your pills. I got the wrong girl."

Several weeks passed. It was not that I was in denial of the pregnancy; I was just simply in the habit of enduring. I was not ready to have a child and I was not quite ready to terminate this potential, either. I did my best to ignore the physical symptoms of pregnancy while I made my way down to Panama, where I e-mailed Mike.

A few days later I was at the San Francisco airport. I waited in line to get my passport stamped, to declare that I was taking no plants or fruits, no fur, no gold, nothing but myself back to this country. Standing on my toes, I peered through the glass plates of the customs booth separating us. I watched Mike looking around for me. I watched him all the way through that line and then I walked right up to him and threw my arms around his leather jacket. He smelled like air and salt and cigarettes, his hair was all messed up. It was late and he'd come straight from work to pick me up. He'd borrowed someone's car. He buried my head in his jacket.

"So nice to see your face," he said.

I could hear my own breath surfacing, the way it seemed to slow for the first time in a month, just knowing there was someone else involved.

His hands were all over my skin. He hoisted my backpack effortlessly onto his back, led me through the airport to the parking lot, never letting go of my hand. Tomorrow we'd talk. Tomorrow, after my breath found its way out of my lungs, after the burden began to shift from me to us.

There had been one particularly bad day of morning sickness in Costa Rica. I'd stood vomiting over a mountain in Rincon de la Vieja National Park on a hike I was about an hour into. Blaming it on the fish or the water, I continued up the mountain, determined to get to the top with time to catch a view. Five hours later, when I got to the top, I saw there was no view, just a cold, blinding fog that made me think of San Francisco. I closed my eyes and hoped for something to happen: for ▶

More Than Just Myself ***(continued)***

the volcano to erupt, for a tourist to appear, for more vomiting. Anything other than the quiet rumbling of what was inside me. There was no sound but the wind whipping against the nylon of my jacket. No lizards or crabs or birds. I cursed the lack of view, the bitter tastes that owned my mouth. But it wasn't the pregnancy that made me long to be anywhere other than there, it was the emptiness of knowing that soon I'd be without the pregnancy. Such a comfort I'd found in being more than just myself.

Coming Closer in the Distance

There is a C. S. Lewis quotation: "We read to know we are not alone." I think the same goes for writing. I wrote for resolution, for the inevitable deepening of self-awareness that comes over the years, and the purpose of bridging the gap between my family and myself. And I wrote because I was isolated from my family and took refuge in the process of recollection and in the company of my characters, who I was better getting to know through writing.

Something happened after the book was released: mornings became clear in a way they'd never been. I felt each day when I opened my eyes that life was beginning over for me. There seemed to be a clearance, permission, to have and do and feel and think as I pleased, and not just in reaction to what I had experienced. The fear that one day, in the middle of it all, I'd wake up and lose my mind completely, dissipated. I was no longer afraid of what lived inside of me.

It was as if I'd stepped into the skin of the person I'd spent ten years trying to become. A protective layer covering the organs of a self that had been destroyed, excavated, rebuilt. It was only after the book was released that I felt I could cover myself entirely in that skin. Breathing became easier, sleep felt lighter. I didn't doubt myself as much. I felt good about what I'd written and put out there. *House Rules* was not the baring of my soul, but the release of all that was holding me hostage to my past. There's a certain amount of diplomacy necessary in writing memoir, in choosing what to include and what to exclude. I tried hard not to cross the line of overexposure. There were several things I left out to protect both my mother and my father. In memoir, the writer must serve as judge in ▶

a court of law, to be somewhat fair and reasonable towards the cast of characters you are representing.

Another thing happened in the release of *House Rules*, something so understated it barely announced itself: I began to feel like I had a place in our family again. It wasn't just accepting or forgiving the past, it was accepting each one of them now, for who they presently are and who they might become. It was less an act of accepting the past and more of releasing it, so that I could step away from my mother and my sister and simultaneously engage them. It's complicated, of course, because both my mother and my sister express their hurt by estranging themselves from me, sometimes for a few weeks, often for months, sometimes years. Engaging them means not giving up on them, trusting, regardless of the patterns of history, that they will come around in their own ways (and in a different way than I have), recognize their part in the way we work, and accept that truth. It was not them I needed to trust, but myself, and in writing *House Rules* I began to trust myself enough to open myself up to my mother and my sister.

For ten years, I held each one of my family members up like a skeleton, trying to disassociate myself from them completely. With the indifference of an archeologist who's found an artifact that's interesting but not what she's looking for, I began to write short stories about the goings-on and events of growing up with my father. It was simply a process of recording memory. I heard my father's voice continuously, every day of my life. His words were like a recurring infection. It was the sameness of his words, and the force with which they hit both at the time when they were intended and later in my memory, after our estrangement, that made the writing come easily. The dialogue in the book, our conversational exchanges, were impossible to forget. They were the same conversations I'd had over again. I've been asked how it's possible that I remember so much, but with the repetitious nature of my father and the concentrated, inescapable bind of our relationship, his words were the easiest material to recall. When I was in graduate school, a professor who had read several of my stories said, "Rachel, do you realize you have no voice in your own writing? Your father completely controls the material. Until you are able to provide some perspective, your father will dominate your story entirely."

Providing perspective required a probing into my current self and our collective past. At first I felt put out by this suggestion. All I'd wanted to do was present my father. He was, after all, a God-given character, and I felt, because it was easier to feel the mightiness of his character than the pain he had caused us, that this man should not go to waste. And that was how I began writing *House Rules*, by building what I later realized was a shrine for my father.

Once this had been dismantled, I began to ask my father questions, questions that he could not directly answer. The question that had plagued me through my childhood: Why had he chosen me?

I've been asked if this was hard to write about, if the memories were painful to reenter. But the most hurtful memories came back with the most ease. They created a sensation my body had memorized and it was the most familiar sensation to return to. This was the space I existed in for the majority of the time I wrote the book. It was only afterwards, having seen it at a bookstore in the airport and paged through it, having never read it cover to cover in its proper book form, when I picked it up and read a passage, that I felt the depth of my sadness. Perhaps it was the timeless space of ▶

being at the airport, the permission and anonymity of that timeless space, but I read the entire passage as if I had not written and rewritten it a hundred times, closed my eyes, and returned the book to its shelf. It was no longer mine.

What would have been more painful than submerging myself in the past is living with it inside of me, taking up the space I needed for love. Writing the memoir was not enough. In writing down the past and letting it go, it was still mine; I owned it, controlled it, and was controlled by it. It was specifically this that was killing me. I no longer wanted to own it and no one else wanted to take it on. It would have been unsatisfying to write and not publish *House Rules*. I needed to get it out of me in order to live my life. It's not just the telling of a story, it is choosing to live life honestly and with integrity. Had I not gotten this story out of me and out into the world, so that it could exist as a separate entity outside of all of us, it would have suffocated my chances of living. I would have imposed that past upon everyone I came into contact with. Perhaps I would have brought the ache of secrecy against the children I might one day have. Had I repressed the experience it would have made me no closer with my family, myself or the world. Simply put, I wrote the book to free myself.

After *House Rules* was published, I received several letters and e-mails from people who'd been affected by my father. Many of them had worked for him at the hospital for over fifteen years. They had their own stories, some more extreme than others. And they thanked me profusely for writing about what was happening at home. They had no idea, like I had no idea of what was going on at work. I'd thought Dad was a hero at work, and to some he was, but to the doctors and nurses and interns and students who knew him well, he'd made their lives "a living hell," in the words that several of them used, and it was vindicating for me to know that there were others. Whether she admits it or not, I read some of these letters to my mom, and I believe it was vindicating for her, too. "So maybe they didn't think I was as crazy as Dad made me seem," she said.

The story is very much alive. The characters are real people. About a month after the book came out, my father responded to it on Amazon. Regardless of the content of his post, this was my father coming forth and speaking his mind. The story did not end with the book; it continues to write itself. This lack of closure particular to memoirs where the characters are still alive is fascinating and at times disconcerting—a reminder that I, as author, have no control over my characters, only an understanding of memory, which I've used to contextualize and better understand my family members, as well as the inevitable curiosity in regards to who we are becoming outside of the text.

Initially I'd hoped that in writing *House Rules* I'd be able to disconnect, to forgive my mother and my father and to walk away from them with the distant fading love of someone who had truly moved on. I thought indifference was a good measurement of my personal growth. I hoped in writing about this, my family would merely be characters in a book. Characters who couldn't hurt me anymore. Characters I could control. Then the book was released and my characters began to react, to offer support, to protest, to call constantly and to stop calling, to offer love and then take that love away. My characters are people, and despite my intent and desire to disengage with them, the publication of *House Rules* brought me closer. It released me from the roll of conductor, shrunk me down to size, and reminded me of what I'd known all my life. ▶

Coming Closer in the Distance *(continued)*

As much as I wanted out, I had a place in this family. Once the book had been published and each of us had reacted, I felt a certain familiar relief—a togetherness with them I'd been afraid to let myself feel for the past ten years, afraid that if I got too close I'd be sucked back into the craziness and lose my potential to live life clear. I prefer being part of something to standing alone and examining it. But the distance was essential. The process of writing *House Rules* was lonely. I can't imagine it any other way, but the publication of *House Rules* was something entirely different. Upon the book's release, I felt closer with my friends, more human, vulnerable. I feel closer with myself.

We're still estranged. I know this makes people uncomfortable, but togetherness doesn't always serve as an indicator of wellness. There's nothing worse than not being yourself, and I couldn't have started becoming myself until I got this out of me. And I couldn't have integrated the self that I was and will always be—the black sheep, the goofy kid, Dad's chosen one—until I came to embrace my mother, father, and sister. We are made up of one another in every way. In making them characters I found the distance I needed to step back and see clearly, to try and understand their intentions and struggles and love. In a way, turning them into characters saved me. And when I was done being saved *House Rules* was published and I could breathe and get on with it. I feel closer to my family than I have since I was a kid, before things went downhill; there is an intimacy in sharing *House Rules* with them, in opening it up and exposing ourselves to the ridicule and compassion of others and each other. I feel like we finally talked, if just a little bit, about what happened. And even if it's a long time from now, if it requires more time for each of us to make sense of our own truths, we'll talk again.

Author's Picks

THIS BOY'S LIFE by Tobias Wolff

Wolff's gentle humor and straightforward approach to his difficult boyhood demonstrate the power and versatility inherent in finding one's own voice.

THE LIARS' CLUB by Mary Karr

Karr's fiery and unapologetic prose, paragraphs that read like poetry, and vivid descriptions rich with meaning showed me the importance of word choice.

A FAN'S NOTES by Frederick Exley

Exley's *A Fan's Notes* is brutal in its honesty and presentation of what it means to be a man. His stream-of-conscious approach to writing freed me up to write what I meant and break away from stylistic constrictions.

ANGELA'S ASHES by Frank McCourt

Frank McCourt has a way of writing that makes the personal appeal universally, interweaving the knowledge of an older man with the simplicity of a child's voice.

DON'T LETS GO TO THE DOGS TONIGHT by Alexandra Fuller

Fuller's brave and honest writing pushed me past the point of worrying over what might offend whom. She gracefully found a balance between revealing details about her family members that were important, uncomfortable, but steadied by love.

STOP-TIME by Frank Conroy

Rather than unfolding events in his memoir in a chronological order, Conroy worked through connected themes, which reshaped the reality of his true stories.

Author's Picks *(continued)*

THE BOYS OF MY YOUTH by Jo Ann Beard

Beard gives her readers permission to laugh at what is not necessarily funny. Her stories, regardless of the subject matter, bring together what's dark and troubling and allow the reader the relief of having a good laugh. In writing *House Rules*, there were several things that I found funny that I knew other people wouldn't. It's important to be able to laugh at what is and what cannot be changed.

SPEAK, MEMORY by Vladimir Nabokov

I remember being particularly moved by the distance with which Nabokov wrote about his brother's death. In describing his death, he revealed so much about his life and their relationship, interrupting the action to draw the characters fully. *Speak, Memory* assigned great responsibility to individual consciousness, the significance of individual memories, which was something I thought about constantly while writing.